1ST Quiz: 25 (Wed.) - OCT. 22

CHAP. 5,6,7.

# STRUCTURAL ENGINEERING

Volume **1**
Introduction to Design
Concepts and Analysis

*Satsop River Bridges* *AISC*

# STRUCTURAL ENGINEERING

## Volume 1
## Introduction to Design Concepts and Analysis

RICHARD N. WHITE

PETER GERGELY

ROBERT G. SEXSMITH

John Wiley & Sons, Inc. New York • London • Sydney • Toronto

Library of Congress Catalogue Card Number: 75-174772

ISBN 0-471-94068-2

Printed in the United States of America.

10 9 8 7 6 5 4 3 2 1

*To Our Parents*

Contents for Series

Volume **1**: Introduction to Design Concepts and Analysis
Volume **2**: Indeterminate Structures
Volume **3**: Behavior of Members and Systems
Volume **4**: Design of Structures

# Preface to the Series

We are convinced that basic courses in structural engineering should emphasize structural behavior and the broader aspects of structures. The full dimensions of the activities of the structural engineer should be portrayed, and proper recognition should be given to the role of automatic computation in our continual search for better designs.

The educational approach used in these books is intended to produce young engineers with a facility for handling new types of structures made of new materials, without a strong dependence on existing design codes and methods of practice. Such engineers will be better suited to undertake a creative and innovative role in the design process—a role that becomes increasingly important and demanding as automated analysis and design capabilities become widely available.

These four volumes present a broader and more unified approach to structural engineering than do most existing publications. A strong emphasis is given to understanding the physical behavior of structures; we feel that the development of this understanding is the first requirement for an imaginative and productive structural engineer. In contrast to the traditional approach in structures, where calculation of forces (called analysis) and determination of individual member sizes (often called design) have been treated as two separate topics, we have arranged the contents of these texts into a more realistic context. The design phase is extended to give more attention to the total design of the entire structural system. Actual engineering projects are utilized in the form of case studies to illustrate the evolution of a structure for a given set of requirements. Analysis is integrated into the design phase, with the details of advanced analysis techniques being given only after the student has become familiar with the many aspects of a project, including preliminary planning, coping with imposed restraints, determining loadings, choosing the structural system or form, and approximate analysis techniques for the system. We feel that this approach is also highly motivational and will transmit to the students some of the excitement and rewards inherent in structural engineering practice.

The philosophy reflected in the scope and organization of the textbooks has evolved in response to a number of developments. The amount of course time available for teaching structural engineering to undergraduates has decreased in recent years, thus placing a high premium on effectiveness in teaching the upper-class courses.

Many civil engineering undergraduates are heading toward careers in water resources, environmental engineering, and other areas where they will not be involved in the actual design of important structures. Flexible curriculums now in force at many colleges will require these students to take only one or two courses in structures instead of the traditional four. We feel that these students should have a background in structures which is quite different from that provided by a first course in analysis or member design. They should understand the planning phases of structural engineering as well as know the fundamentals of structural behavior and equilibrium analysis of structural systems. An encyclopedic treatment of structural analysis and design is not desirable in books to be used by undergraduates.

Finally, we observe that routine analysis and design situations are now being handled quite well with the aid of digital computers. This development is bound to proliferate in the years ahead. The structural engineer of the future can best devote his energies and talents to decisions on structural form and the control and understanding of behavior, thus extending his concern beyond detailed analysis and design procedures.

The organization of the material in the four volumes reflects our response to these observations. The following summary of material coverage is amplified in individual prefaces to each of the volumes.

Volume 1 sets the activity of structural engineering in context through consideration of the case study of a bridge design. This is followed by a discussion of structural requirements and loads, structural form, analysis of determinate structures, and approximate analysis of indeterminate structures.

Volume 2 consists mainly of a rigorous presentation of indeterminate analysis techniques. Analysis methods covered include the virtual work method, moment distribution, force and displacement matrix analysis approaches, and nonlinear analysis. The role of analysis in the design cycle is also explored at some length.

Volume 3 is devoted to an in-depth treatment of the behavior of a wide variety of structural components and systems, subjected to loadings ranging from working load up to ultimate capacity. Fundamentals of behavior and design of steel, concrete, and timber structures are presented largely independent of any design codes or specifications.

Finally, Volume 4 treats design in accordance with the provisions of the latest editions of the pertinent U.S. building codes and specifications. Design of the actual structures introduced in the other volumes is supplemented by additional project-oriented design. This book is the only one of the series that is strongly dependent on code provisions.

## Use of the Books

The four volumes are arranged to facilitate usage of the books for a variety of possible course combinations, ranging from a single course for the nonstructures major to a full complement of four courses. Volume 1 is considered essential in any sequence. Although some of the early material can be assigned as reading without requiring classroom time, most instructors could profitably weave in their own experiences and further capitalize on the motivational value of this volume.

Suggested coverages for different course combinations are given below. The standard course unit is defined as a three-hour or credit, semester-long course. It is assumed that the student will have completed courses in statics and mechanics of deformable solids prior to studying structures.

One possible instruction method involves the mixture of analysis, structural behavior, and design in a more or less continuous blend through the first three or four courses of structural engineering. This approach has been used for a number of years at Cornell University and is certainly favored by the authors as the optimum manner for teaching structures. The second, more traditional, approach divides material into analysis and design, most often with determinate and indeterminate analysis comprising the first two courses and steel and concrete design the last two. This approach may result in a misconception of the role of the structural engineer on the part of those who elect only one or two of the courses in such a sequence. Although we feel that this method of presenting structures tends to separate too widely the various aspects of structures, we also realize that there are organizational conveniences to this mode. Suggested textbook coverages for both methods are given below.

1. *A mixture of design concepts, analysis, and behavior,* with the possibility of a student terminating his instruction in structures after any of the courses are completed.

(a) One course—we feel that a single, terminal course in structures should emphasize a broad introduction to structures, equilibrium analysis of statically determinate and simpler forms of statically indeterminate structures, and some basic aspects of behavior of either steel or reinforced concrete structures. Statically indeterminate analysis is restricted to approximate analysis methods. Elementary behavior of both steel and concrete structures should be covered if the course credit is four hours instead of three.

This course would use the entire contents of Volume 1 and selected portions of Volume 3.

(b) Two courses—some institutions require two structures courses while others have the second course as an elective. The division of material for the first alternative may well be different than with the second, but the total coverage at the end of the two courses would include that given above for the single course, plus displacement analysis, an introduction to indeterminate analysis using virtual work, additional work in equilibrium analysis, and continuation of study of structural behavior.

This course sequence would use the entire contents of Volume 1, most of Volume 3, and a portion of Volume 2, including displacement analysis by the methods of moment area and virtual work, and extension of displacement analysis to statically indeterminate analysis by the force method.

(c) Three courses—the two courses described above would be followed by a third course that is devoted mainly to analysis of indeterminate structural systems. Force and displacement methods and computer-based matrix analysis would be covered in depth, and an introduction to collapse theory and plastic design would be included. The analysis would be coupled with selected design problems. Finally, the course would be used to further the student's overall concept of structural action and behavior.

The entire contents of Volumes 1 to 3, and part of Volume 4 would be covered at the termination of a three-course sequence.

(d) Four courses—the fourth course is a design course that integrates the material covered earlier in direct application to practice-oriented design problems. The design of both structural systems (or forms) and structural elements would be treated through the mechanism of short problems and a comprehensive design project in which teams of students proceed through the entire design process, including oral presentation of preliminary and final reports to a faculty and student jury.

2. *Sequential approach; analysis followed by behavior and design.*

(a) First course—basic analysis methods. Use Volume 1 and the first three chapters of Volume 2.

(b) Second course—indeterminate structural analysis. Use the remainder of Volume 2.

(c) Third and fourth courses—behavior and design of steel and reinforced concrete structures; one course for each. Use half of Volumes 3 and 4 for each course.

There are other combinations of course content that may be preferable to a given instructor in a given situation. We planned the division of material in the four volumes to encourage innovation and experimentation in teaching structural engineering to undergraduates. The fourth volume is dependent upon the third, but any of the first three could be used by themselves.

No matter how these volumes are used, their major characteristic is a constant emphasis on a full understanding of how structures actually behave.

Chapters are numbered consecutively through the four volumes. For example, Volume 2 begins with Chapter 8. The Table of Contents for the Series, included in each volume, provides a key to the appropriate volume for any given chapter. Sections, equations, examples, and problems are numbered sequentially in each chapter (for instance, 3.1, 3.2).

Richard N. White
Peter Gergely
Robert G. Sexsmith

Ithaca, New York
November 1971

# Preface to Volume 1

This volume contains extensive material on the many phases of structural engineering, including conceptual planning, evolution of structural systems, determination of loadings, preliminary design and analysis, and evaluation of alternative structural systems. It covers analysis of statically determinate structures and approximate analysis of statically indeterminate structures, with an emphasis on understanding the behavior of the entire structural system under analysis.

The first chapter is a case history study of the engineering involved in the creation, design, and construction of the Solleks River Bridge in the state of Washington. The general activities of a structural engineer in a broader context than bridge design are also illustrated.

The objectives of structural design are treated in Chapter 2. An introduction to what we might call "structural philosophy" is put forward. Analysis is placed in its proper perspective and important topics such as safety, serviceability, and design codes are considered.

The character and intensities of common loadings are described in Chapter 3. The presentation is designed to encourage critical thinking on the part of the reader, particularly with regard to the variabilities and uncertainties involved in determining the loadings for a particular structure.

Structural form is covered in Chapter 4, beginning with simple forms that carry applied loads by pure tension or compression, proceeding through mixed tension-compression structures, bending forms, and flat and curved surface structures. Relationships between natural and man-made structures are explored.

The fundamental concepts basic to analysis are presented in Chapter 5. These concepts include elasticity, linearity and nonlinearity, superposition, geometric instability, and determinacy of structural systems. This material is considered to be an essential prelude to numerical calculations.

Chapter 6 covers the analysis of determinate structures; it is the longest chapter in the book. The approach to analysis stresses the uniformity of equilibrium analysis for all types of determinate structures. Tension structures are treated first, followed by analysis of other forms in the same order as in Chapter 4.

The final chapter applies equilibrium analysis methods to the approximate analysis of statically indeterminate structures. Results are compared with conventional analyses, and the reader is shown that quick approximate analyses of highly complex structures can be performed.

An extensive set of problems, consistent with the philosophy followed in this book, is included. Answers are given for about half of the problems. We urge the instructor to utilize the problems early in the first course, and to supplement them with similar problems drawn from his own experiences. Assignments from the suggested reading lists are also encouraged. Far too few students make good usage of the rich storehouse of engineering material available in their libraries.

Finally, some laboratory exercises are described in the book. The use of these exercises and other physical teaching aids can enrich a course both technically and motivationally. They should be employed whenever the local laboratory facilities permit.

Richard N. White
Peter Gergely
Robert G. Sexsmith

Ithaca, N.Y.
November 1971

# Acknowledgments

We are grateful to William J. Hall (University of Illinois), Halvard Birkeland (ABAM Engineers, Inc.), J. W. N. Fead (Colorado State University), and Dean S. C. Hollister (Cornell University) for their careful reviews and constructive comments. Their suggestions have been invaluable in preparing this book. We especially thank Halvard Birkeland and his associates at ABAM Engineers, Inc., for providing materials and information for the case study presented in Chapter 1. We are also indebted to the many organizations and individuals who have willingly supplied photographs and illustrations.

We acknowledge the unseen but essential contributions of graduate and undergraduate students for their assistance in checking examples and problems, and the several typists who worked on the various drafts of the manuscript. Special thanks are extended to the publisher's editorial and production staff for their enthusiastic assistance and support in all phases of this project.

Finally, we thank George Winter and our other colleagues at Cornell University for helping to create the imaginative teaching philosophy and educational outlook that stimulated us to write this series of textbooks.

R. N. W.
P. G.
R. G. S.

# Contents

## Volume 1

# STRUCTURAL ENGINEERING

Volume **1**

Introduction to Design Concepts and Analysis

# CHAPTER 1

# The Evolution of a Structure

Solleks River Bridge, Washington. *ABAM Engineers, Inc.*

*The Scientist explores what is.*
*The Engineer creates what has not been.*
Theodore von Kármán

# 1 The Evolution of a Structure

What is structural engineering? What does the structural engineer do? How does the work of the structural engineer relate to that of a highway designer, a water resources engineer, or an aeronautical engineer? The answers to these questions are neither simple nor unique because of the broad scope of structural engineering. We find that buildings, bridges, dams, transportation systems, water and sewage treatment facilities, aircraft frames, power generating stations, storage tanks, and many other types of constructed facilities have a common denominator—a structural system that must have adequate strength to resist safely the many loadings that act on it during its lifetime. The engineering of this wide variety of structural systems is the province of the structural engineer.

We shall examine the functions and responsibilities of the structural engineer through the mechanism of a case study by presenting a summary of the engineering and decision making leading up to the construction of an actual bridge in the state of Washington. The case study will illustrate the fundamental considerations involved in arriving at an engineering solution to a real situation. It will be followed by a broader discussion of the scope of structural engineering.

## 1.1 SOLLEKS RIVER BRIDGE—A CASE STUDY

The Solleks River Bridge spans a 150-foot deep river canyon in the Olympic Peninsula in northwest Washington. The purpose of the bridge is to provide an access road for logging operations in the state-owned forests. Hemlock, cedar, spruce, and Douglas fir timber are being harvested on a perpetual yield basis under the management of the Washington Department of Natural Resources.

The bridge is located in extremely rugged terrain; the site is accessible only by an 85-mile trip from the city of Aberdeen, and many miles of the trip are on rough roads. The area is located in the Olympic rain forest, where about 140 inches of rain falls each year and the temperature ranges from about 20 to 90 degrees F.

The owner, or client, for the bridge construction was the state of Washington. The owner's requirements were quite simple and included the following ones.

1. The bridge must be able to carry extra-wide, heavy logging and construction equipment as well as off-highway logging trucks.
2. The bridge must be rugged, have low initial cost and low maintenance charges, and have a long life.

The final solution to this set of requirements is shown at the front of this chapter. Designed by the Tacoma consulting firm of Anderson, Birkeland, Anderson, and Mast (ABAM), the bridge received a Prestressed Concrete Institute Award in 1969 for excellence in architectural and engineering design using prestressed concrete. We shall seek to re-create some of the strategies and decisions that led to the adoption of this particular design. Prior to this, however, we should consider the basic steps involved in the evolution of a structure from the first concepts of need to the placing of the final element in the completed structure.

Engineering design can be summarized as a five-step process:

1. Recognizing needs and specifying objectives (general planning).
2. Preliminary design of alternate solutions.
3. Evaluation of alternatives (analysis).
4. Final design and analysis.
5. Implementation.

We shall illustrate the many facets of each of these steps as met in the design of the Solleks River Bridge. It will become evident that the designer does not move directly through these steps, finishing one completely before he moves on to the next; instead, he is constantly looking ahead as well as looping back to reexamine his earlier work. We should also note that the details of the design process are unique for a particular project. A multistory building design would follow the general five-step process, but the relative time spent on each of the steps might be much different than for the design of a highway bridge.

## General Planning

An engineering project begins with the recognition of need and the establishment of design objectives. The general scope of the facilities needed to satisfy the requirements is then set down. Project financing is explored and a justification for the planned facility is sought.

In our case study the need for the bridge can be considered from two levels—the continued general demand for utilizing our natural resources, and from direct decisions of the Washington State Department of Natural Resources.

The department wished to make accessible a 10,000-acre tract of state-owned land in order to manage and improve this natural resource of the state of Washington. Coupled with the road and bridge is a 100-man honor camp facility that provides a continuing work force for the reforestation and management activities of the department. Financing for the honor camp, and implicit approval of the venture, was obtained in part by the successful passage of a statewide referendum in November 1966.

The economic justification for the bridge on this rather modest public project was thus made by a governmental agency; the process was quite straightforward compared to the lengthy and complex questions that must be answered prior to going ahead with the construction of such facilities as the $300 million Verrazano Narrows suspension bridge or the $750 million World Trade Center buildings in New York City. Nevertheless, we can be sure that the decision to build the bridge and open up the timberland was not by any means unanimous among the people of Washington; conservation groups interested in preserving large tracts of untouched forest land for future generations have received increasing support in recent years. The question of the value of a managed, selectively harvested timber area versus a completely untouched virgin forest does not have a single "correct" answer. Many developments in the public domain, particularly those affecting our environment and manner of living, require some rather painful decisions. Engineers should be prepared to assert personal as well as professional opinions in these cases.

A minimum cost bridge was sought. Getting the most for the client's dollar is one of the unwritten laws of engineering. In fact, an engineer has been defined as someone who can do for one dollar what any fool could do for three dollars. The economic justification for a project that must be financed with borrowed money, such as a toll highway or a large office building, is a complex and crucial undertaking. Both the project costs and a monetary value of the benefits must be established. If the cost-benefit ratio is not favorable, the entire venture may be dropped. During these economic studies the engineer is asked to provide accurate projected estimates on the cost of the proposed facility. Cost estimates are best done by engineers with considerable experience who work either with figures from similar completed projects or from approximate designs.

## Preliminary Design of Alternate Solutions

This part of the design phase is perhaps the most important of all in that the success of the final structural system is directly dependent on the preliminary design activities. These activities include proposing a number of structural systems (forms) that seem suitable for the situation. This "idea" stage is followed by a determination of approximate values of the major design parameters for each feasible alternate. The calculation of approximate forces in each structure is a prerequisite to establishing these design parameters (such as the depth of a beam, the overall dimensions of a column, or the size of a foundation).

The creativity and imagination of the structural engineer are paramount attributes in the preliminary design phase. An understanding of construction procedures is also essential to ensure that the proposed alternates can be built satisfactorily and at a predictable cost.

The Solleks River Bridge had to span over 200 feet across a 150-foot deep canyon, supporting a 75-ton logging truck as well as its own weight (dead load). One side of the canyon has overhanging cliffs and the other side is almost as rugged. The site is in a remote area with no local labor, concrete mixing facilities, or other construction aids. Choice of construction materials is particularly important in this type of site.

The bridge is subjected to three primary loadings: dead load, truck loading, and wind acting on the side of the bridge. Other loadings include earthquake forces, the effect of truck braking, and temperature and other environmental effects.

The only structural system (form) that received serious consideration by the ABAM engineers is shown in Figure 1.1*. The evolution of this form for the bridge was influenced strongly by the geometry of the steep canyon crossing. The canyon is excessively deep for any type of central vertical support; also, a support in the river could lead to many additional problems such as producing possible obstructions to flow during heavy run-off periods as well as affecting the fish life. An alternate solution

*The idealized structural system in Figure 1.1*b* consists of a line diagram with appropriate external reactions and internal hinges. The lines represent the centerlines of the various members. Internal hinges are shown as open circles (—o—) and are assumed to be frictionless, transmitting only concentrated forces from one segment of the structure to another. External reactions are of two types here—hinges ( or ) and rollers ( or ). The first type can supply a force reaction in any direction while the second can resist only those loads normal to the plane on which the roller rests. An actual hinged support for a bridge girder is shown in Figure 1.2.

Another common reaction is the fixed support ( ), which can resist a force in any direction as well as a bending moment.

We shall utilize these conventions throughout the text.

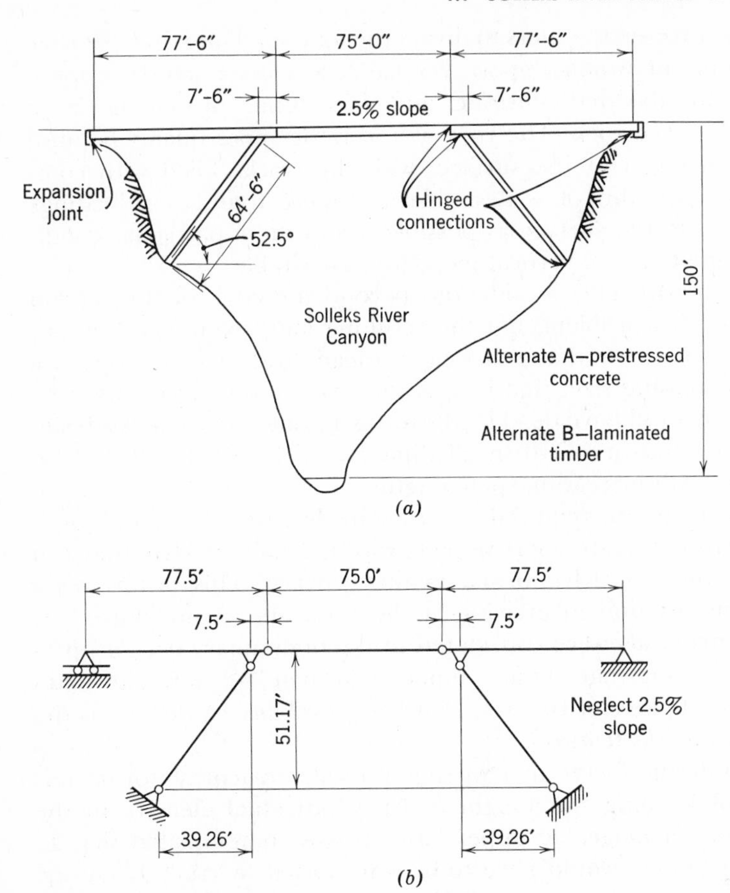


**Fig. 1.1**
Forms considered in preliminary design. (*a*) Elevation view of proposed bridge. (*b*) Idealized structural system.

**Fig. 1.2**
Hinged support for a bridge girder.

*Bethlehem Steel*

would be to use two vertical piers to divide the total span into three shorter spans. This solution would impose vertical forces on a steeply sloping formation of interbedded siltstone and sandstone, producing large shearing forces in the rock. The rock was of rather poor quality because of severe fracturing near the surface, with the cracks filled with compacted clay and silt; thus it is desirable to impose foundation loadings as nearly normal to the surface as possible. From these particular conditions we conclude that two vertical piers are undesirable.

A simple span with end foundations beyond the edge of the canyon would eliminate this problem, but the resulting long span would be uneconomical. We observe that self-weight (dead load) stresses increase linearly with increasing size; that is, a beam carrying only its own weight on a 200-foot span will have dead load stresses 10 times as great as a beam scaled down by a factor of 10 in all dimensions. Live load effects also increase rapidly with increasing span length.

Intermediate piers are required in order to decrease the span length of the bridge girders. If the piers slope toward the rock surface, they can impose a load more nearly normal to that surface. This factor was a paramount consideration in arriving at the form shown in Figure 1.1. Two different materials were considered in the preliminary design—prestressed concrete (alternate A) and timber (alternate B). Both alternates use a minimum of site-cast concrete. The inaccessibility of the site is the obvious reason for this feature.

Another significant factor in choosing possible structural forms was the necessity of keeping the weight of each individual element in the structure to some manageable upper limit. It was contemplated that all elements of the bridge would have to be transported by truck from one of the nearest metropolitan areas (155 to 210 miles away). Finally, the great difficulty in gaining access across the canyon and down the canyon walls was a decisive factor in choosing possible structural forms.

We shall concentrate our discussion on the approximate analysis and determination of major design parameters for alternate A (Figure 1.1). The structural concept is elegant in its simplicity. Prestressed concrete girders rest on foundations on each side of the canyon and are supported near their ends by inclined legs that transmit the inner beam reactions to the rock walls of the canyon. The central span is supported from the cantilevering portions of the two end spans. Initially the girders are connected by hinges, but later in the construction phase they are given continuity and bending resistance at the connection in order to resist the effects of the truck loading more efficiently. Three girders as shown in Figure 1.3 and three pairs of struts are proposed in the cross section of the structure.

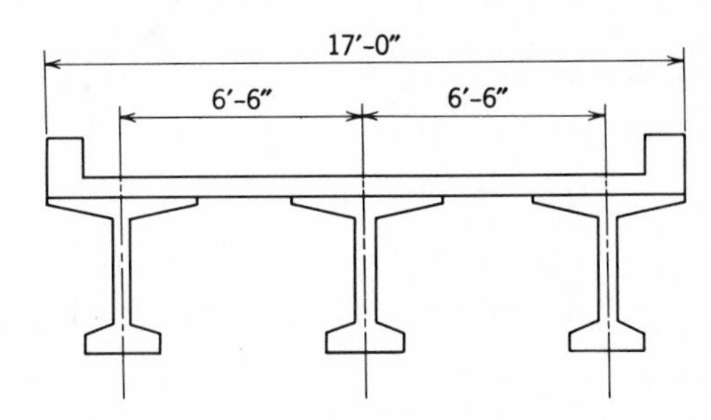


**Fig. 1.3**
Geometry of alternate A.

(*a*) Preliminary cross section, alternate A.

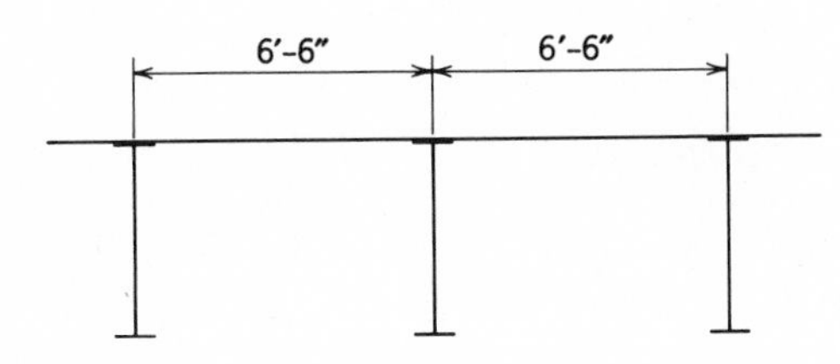


(*b*) Idealized preliminary cross section.

Prior to making the girders continuous at the connections, this structural system is statically determinate; that is, the magnitude and distribution of forces in the main members of the bridge can be determined by application of the equations of static equilibrium (see Chapter 5). In a statically determinate structure, the forces produced by the live loading are independent of the size of the various members. Of course, the self-weight stresses are a direct function of the size of the structural members. They must be arrived at by trial and error in most structures.

In the trial-and-error procedure the engineer estimates dead loads. Member sizes required to resist the dead load and other loadings are first determined. The actual dead loads are then computed and compared with the estimate. Several cycles may be necessary to get high accuracy, but usually one cycle of calculations is sufficient in the preliminary design stage. External reactions and approximate bending moments in the structure are shown in Figure 1.4 for an approximate dead load of 1.0 kips/ft for one girder and its share of the other concrete elements, and 0.6 kips/ft for the supporting strut. The dead loads are computed using unit weights of 120 pcf (pounds per cubic foot) for the precast, lightweight concrete girders and struts, and 150 pcf for the cast-in-place deck slab, curbs, and diaphragms. The diaphragms are transverse vertical slabs spanning between the girders at several locations along the bridge.

The structure becomes statically indeterminate after the main beams in the bridge are made continuous at the interior connections. Analysis of statically indeterminate structures requires consideration of structural

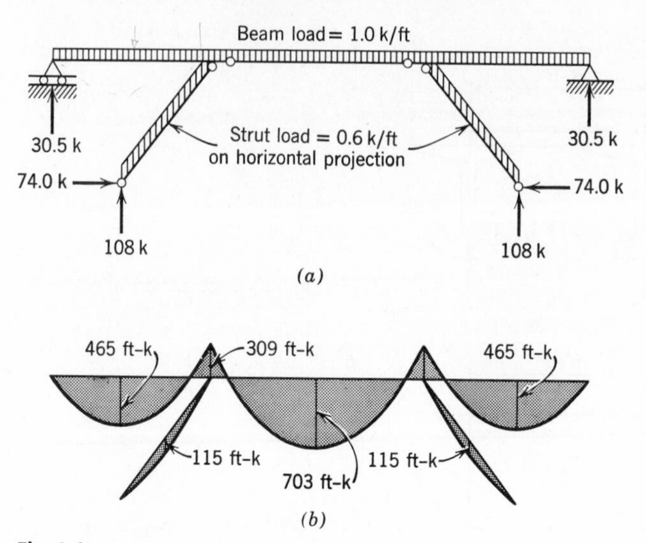


**Fig. 1.4**
Approximate dead load reactions and moments. (*a*) Approximate reactions for dead load. (*b*) Bending moment diagram for dead load.

deformations in addition to statics. The distribution of forces in the structure are a function of the sizes of the members. The designer's dilemma is that he must determine forces in order to pick member sizes, but the magnitude of the forces will be influenced by the sizes he picks!

A good solution to this dilemma is to utilize approximate analysis techniques for the first analysis. Sufficient assumptions are made to transform the indeterminate structure into one that can be analyzed by statics. For the Solleks River Bridge, the approximate analysis includes making two assumptions as to the location of changes in curvature in the continuous beam span. Approximate analysis is discussed in Chapter 7; it can be carried out quickly and will yield forces of reasonable accuracy for the preliminary design phase.

In addition to gravity loadings, the bridge must resist horizontal loads produced by wind and possibly earthquake.

The Solleks River Bridge is loaded with extremely heavy logging trucks, logging equipment, and construction equipment needed for later projects. The logging truck load is critical here; its design weight of 75 tons is distributed according to the wheel pattern shown in Figure 1.5*a*. This type of truck is not permitted to travel on regular highways; it is more than twice as heavy as the normal truck loadings described in Chapter 3.

Approximate external reactions and girder bending moment for the truck located with its center of gravity in the middle of the bridge are

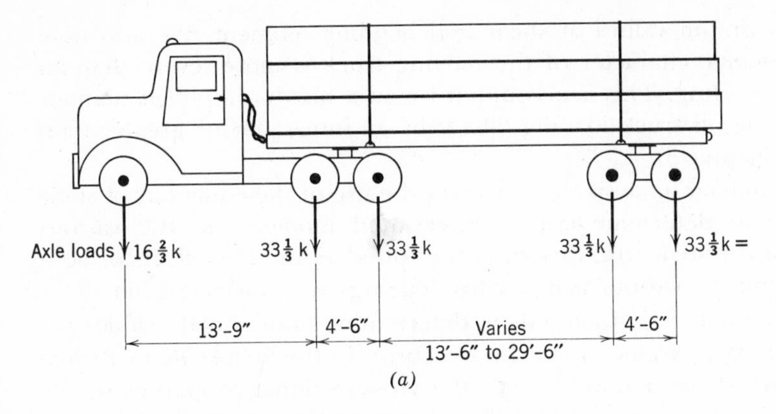


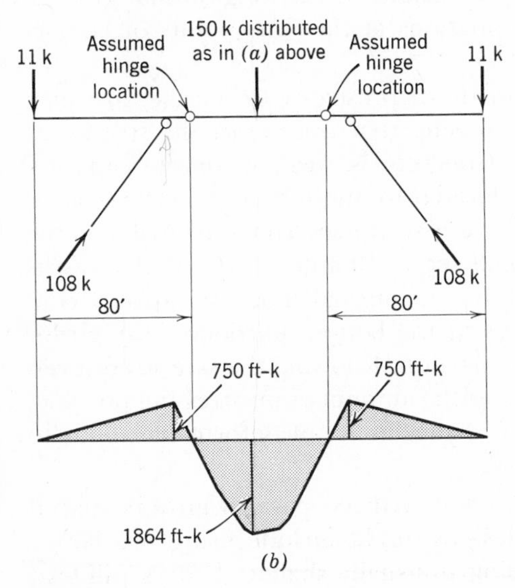


**Fig. 1.5**
Live load on bridge. (*a*) Design truck loading. (*b*) Reactions and bending moment with truck at center of bridge.

shown in Figure 1.5*b*. Note that the indeterminate beam has been made determinate by assuming the location of points of zero moment in the center span. This loading case is only one of several that the designer would consider in the preliminary design, since the effects of the truck traversing the entire length of the bridge must be known. Some portions of the structure will be more highly loaded when the truck is near the end of the span. The analysis of structures subjected to moving loads is treated in Volume 2, Chapter 16, where techniques are given for finding the absolute maximum values of shear and bending moment. We also note that the dynamic character of the moving truck is more severe than an equal static loading. This is accounted for in a highly simplified manner by multiplying all truck loading effects by an impact factor greater than unity (see Chapter 3).

The preliminary analysis thus involves the use of the equations of static equilibrium to determine critical forces and moments in the various members of the structure, including the foundations. The effects of both dead load and the various primary live loadings are considered here.

The force analysis is followed by determining approximate values for the major design parameters of the structure. In the Solleks River Bridge (alternate A), these parameters are the cross-sectional properties of the supporting struts, the shape and dimensions of the longitudinal girders, and the type and size of the foundations at the four points of contact with the canyon.

Alternate A shows the girders made of prestressed concrete, in which forces are introduced by the prestressing that counteract the stresses as a result of the applied loadings. Concrete is weak in tensile capacity; thus the prestressing elements are located to minimize the tensile stresses in the concrete. How is the size of the prestressed concrete girder in the center span determined? As we shall see in Chapter 4, the girder resists the effects of the applied loading by an internal moment capacity consisting of resultants of tensile stress in the bottom portion of the girder and compressive stress in the upper portion. Knowing the safe stress levels of the materials used in the beam, and the amount of prestressing possible, the engineer applies the principles of mechanics of deformable solids to arrive at the appropriate girder size.

The sectional view in Figure 1.3 shows three parallel girders spaced 6′–6″ on centers. Other feasible designs might include using two larger or four or five smaller girders. Using unusually shallow girders will lead to excessive deflections and "bounciness" as the truck crosses the bridge, however, and a minimum girder depth can be established on the basis of allowable deflection under live load. We should note that occasionally the designer must use lightly stressed members in order to control the magnitude of service load deflections.

A reinforced concrete deck slab is cast in place after the beams are erected. This slab acts as a top flange for the girders and helps resist the truck loading. It also spans from one side of the canyon to the other as a horizontal beam, resisting wind and earthquake forces.

The strut cross section and the foundations are also sized in the preliminary design phase. The characteristics of the girder and strut elements will be discussed further in Chapter 4 where we consider the efficiencies of various structural forms. The calculations leading to the approximate design are also given in subsequent chapters.

## Evaluation of Alternatives

Evaluation of the proposed alternative solutions is usually going on at the same time as the preliminary design. Some proposals fall by the wayside simply because they look so unfavorable in comparison to the better alternates. In many situations, however, we find a number of alternates of nearly equal merit. Then we must assess the suitability of the alternates in relation to other requirements in the project, compare construction problems, critically examine overall structural efficiency, and consider the economics and esthetics of each alternate.

Alternate A was chosen as the structure best suited to the requirements set for the Solleks River Bridge even though its estimated cost was slightly more than that of alternate B, the timber structure. Since both contenders for the final design utilized similar geometry, the choice was between construction materials.

One may wonder why alternate B, the timber structure, was not used for a logging road bridge, particularly since its estimated cost was less than for alternate A. The answer lies in problems related to the bridge deck. It was felt that a timber deck would not wear as well as concrete when subjected to the severe combination of heavy wheel loads and abrasive action of coarse gravel pulled onto the bridge surface by the logging trucks. A concrete deck on timber beams is not a good solution here because of the difficulties in fastening the two materials together. A high quality connection is essential for this bridge because of the continual pounding action of the logging trucks.

The timber alternate utilized glued laminated members, in which many boards of small dimension are glued together to make up a beam. The process produces large timber members of higher quality than a natural timber made from a large tree. It has given a substantial impetus to the use of timber members in heavy construction.

The question of prestressed concrete versus structural steel is quite often dependent on local circumstances. Concrete requires no paint and accordingly is easier to maintain than most structural steels. The requirement for low maintenance, along with the excellent prestressed concrete plant facilities in the Pacific Northwest, gave the edge to prestressed concrete for the Solleks River Bridge.

The prestressed girder dead load is minimized by utilizing a lightweight aggregate instead of normal rock in the concrete. The same lightweight concrete, which weighs about 25% less than regular concrete, is used in the struts. The struts also have a hollow cross section to further reduce their dead weight. The high efficiencies of both the beams and the compression struts are revealed in more detail in Chapter 4. The reactions and the internal forces are calculated in Chapter 6.

Expected difficulties in construction were minimized by constant attention to the construction process by the designing engineers. As an example, the upper limit on girder length was set at about 78 feet since the girders had to be transported to the site over a narrow road with tight curves. Conversations were held with prospective builders throughout the design process. This type of planning is difficult to accomplish in any country that uses competitive bidding to choose the contractor for the construction; the designers do not know which firm will be building their structures until after the plans and specifications are completed and construction bids are received, opened, and evaluated.

### Final Design and Analysis

The final properties of each member and connection in a structure must be determined after a choice of system has been made. The member sizes arrived at in the preliminary design phase may be satisfactory for final dimensions; if not, they are used as a starting point for the final design. The refined analysis will be done with a digital computer if the structure is complex. Even some of the final selection of member sizes may be done with the aid of a computer.

Final design dimensions for the prestressed girders and struts for the Solleks River Bridge are shown in Figure 1.6. The profile of the internal

**Fig. 1.6**
Final design of Solleks River Bridge.

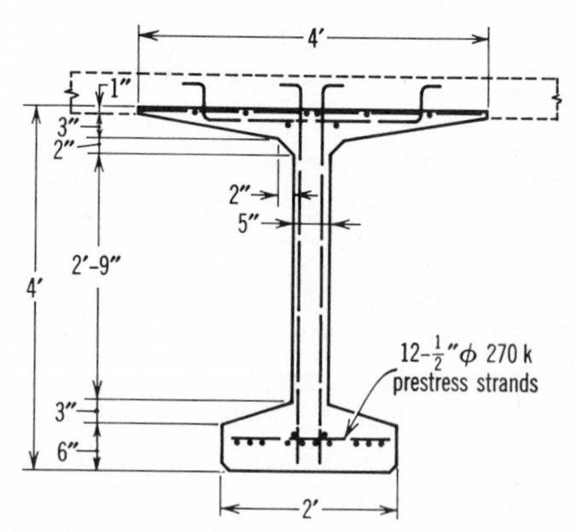


(a) Girder section; girders and 5-in., cast-in-place deck act compositely.

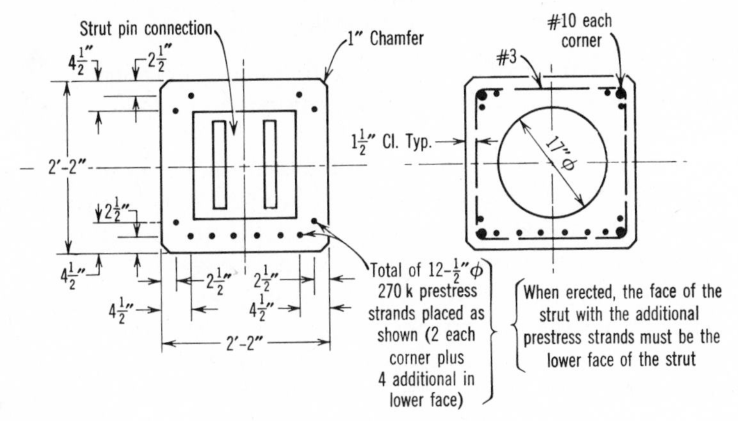


(*b*) Strut cross section.

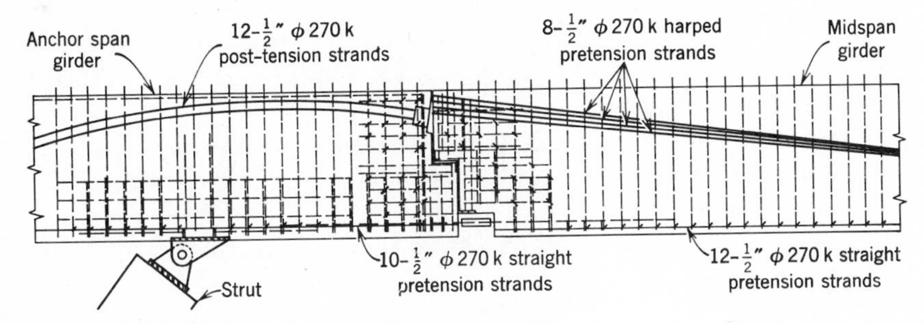


(*c*) Prestressing element profile and girder connection detail.

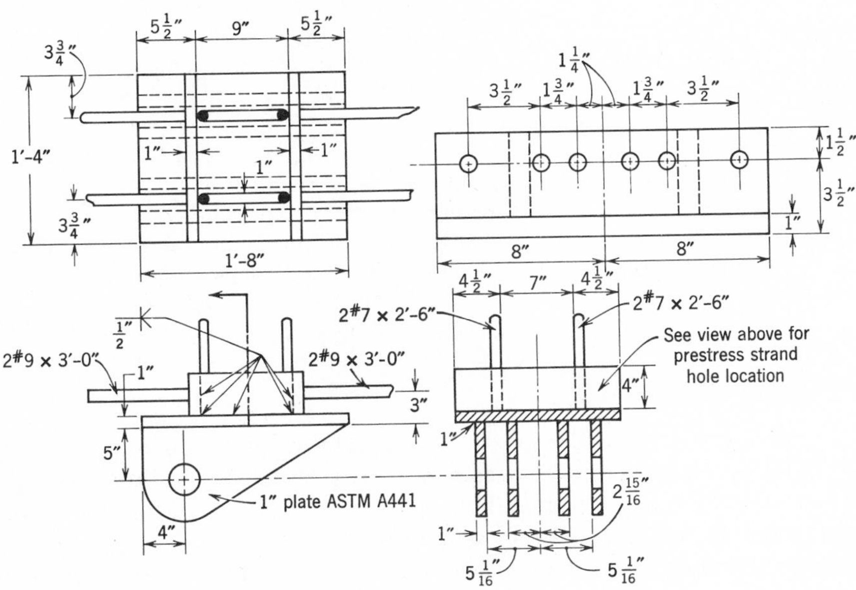


(*d*) Connection detail—girder to strut.

prestressing element in the vicinity of the strut-to-girder connection is also shown; the determination of this profile is one of the more difficult tasks in this design.

The connections in a structural system are vital in ensuring integrity of the system. A typical hinged connection for the Solleks River Bridge is illustrated in Figure 1.6*d*. We shall investigate the design of this connection and other common structural connections in Volume 4.

As the structural engineer develops his design calculations and sketches, he presents his results to draftsmen who prepare drawings of the entire structural system and its members and connections. Close coordination between engineer and draftsman is needed to ensure a satisfactory set of final design drawings. The drawings are an absolutely essential communication channel between designer and builder for the fabrication (manufacture) of the many components in the structure and for the construction process itself. The design documents for the Solleks River Bridge included nine pages of drawings.

The fabricating firms normally prepare additional drawings that show every detail such as locations and sizes of all steel and hardware. The design engineer is responsible for checking the accuracy of the fabrication drawings and occasionally for checking special construction drawings prepared by the builder.

The drawings are supplemented by written specifications that spell out in detail the scope of the work, the materials to be used, the required quality of workmanship, the applicable standard codes, etc. Fifteen pages of technical specifications were prepared for the Solleks River Bridge. Below is a sample clause of the section on construction details for cast-in-place concrete.

> Due to the remoteness of this project and the time necessary to haul ready-mixed concrete to the site, cement shall be added to the batch trucks at the site. As an alternate, a concrete batch plant may be set up at the bridge site or within a distance of the bridge site to be able to place concrete within the time specified in the Standard Specifications. The method selected shall be subject to approval by the Engineer before use.

## Implementation

The final and most visible phase of structural engineering is the implementation of the design. This construction phase is the culmination of the aspirations of the client and the efforts of the engineer; the design documents are transformed into reality.

The amount of control exercised over the construction phase by the design engineer varies considerably from one project to the next. For our case study structure the designers were involved in construction planning and problems from the outset of the project. As an example, the bridge members were placed out over the canyon with the support of a cableway constructed across the canyon. ABAM engineers investigated several factors, such as the maximum cable size normally available in a logging operation and the maximum cable sag feasible at the site, in their determination of the limiting weights and sizes for bridge members. They also specified the sequence of work operations for the bridge.

A. Excavate rock for footings and abutments.
B. Drill and grout anchorages for footings and abutments.
C. Cast footings and abutments (not including the back and wing walls of the abutments).
D. Prestress the abutments to the rock canyon wall and grout the prestress tendons.
E. Swing each unit consisting of abutment section of a girder and strut into place on the footing and abutment and weld the 4—#10 rebar anchors to each girder hinge. It is anticipated additional temporary bracing will be required.
F. Cast the end diaphragms and diaphragm numbers 1, 2, 5 and 6.
G. Erect central span girder sections and grout compression pads.
H. Cast diaphragms number 3 and 4 and cast roadway slab.
I. Weld bottom reinforcement and drypack girder joints.
J. Cut the 4 — #10 rebar anchors to each girder hinge at the north abutment (expansion abutment) and cast the anchor blocks at south abutment.
K. Remove anchorage hardware at north abutment (expansion abutment) and cast wing and back walls.

The engineer responsible for the design of a structure may also be responsible for seeing that it is built according to the plans and specifications. Often he appoints a supervising engineer who represents him and stays at the site. He may find it necessary to prepare revisions to the design as the construction phase reveals new knowledge about the situation. This often happens with the design of foundations since it is impossible to know the entire picture on the local soil or rock until the excavations are completed.

Early work at the Solleks River Bridge site revealed extensive weathering and fissuring of the canyon wall rock. ABAM engineers were called in by their client to furnish the additional engineering required for the safe construction of the bridge.

Finally, the client and the designing engineer act together to minimize disruptions of the environment as produced by the construction. We find the following provisions in the Special Conditions section of the Solleks River Bridge documents.

A. No heavy equipment shall operate within the wetted perimeter of the stream.
B. Equipment crossings shall be held to a minimum and be confined to one area.
C. Extreme care shall be exercised to insure that no fresh cement, petroleum products, chemicals or other deleterious materials shall enter the stream.
D. These operations are to be controlled to allow a very minimum of siltation to the stream.
E. Under no circumstance is there to be created a block to stream flow or fish passage.
F. Debris from this project is to be removed and disposed of by burning or placing beyond high water flows.

The Department of Fisheries and the Department of Game reserve the right to make further restrictions if deemed necessary for the protection of fish life.

## 1.2 THE SCOPE OF STRUCTURAL ENGINEERING

We have explored the role of the structural engineer in planning, designing, and building the Solleks River Bridge. In order to better define the broad scope of structural engineering, we shall seek answers to two questions: What other types of facilities fall within the province of the structural engineer, and how are his activities and responsibilities different for each type of structure?

The structural engineer is involved in the design of hundreds of types of structures, including those used in bridges, buildings, industrial plants, dams, water and sewage treatment facilities, water tanks, power stations, sports facilities, pressure vessels, airplanes, spacecraft, antennas, nuclear reactors, subways, tunnels, pipelines, transmission towers, and ships. The structural systems used in several of these types of facilities are shown in Figure 1.7; they range from the continuous, doubly-curved concrete surface of a modern high arch dam to the wispy, delicate assemblage of small linear elements in an antenna structure.

The structural engineer interacts almost continuously with other engineers and with architects. In many commercial building designs (office buildings, high-rise apartments, university complexes, convention centers, and similar facilities), the architect has the primary responsibility. He assesses the needs of the client and evolves a design of space and environment that meets those needs. Acting in conjunction with the structural engineer, he develops the overall shape and appearance of the building. Several alternate schemes are normally considered.

**Fig. 1.7** (a)
Flaming Gorge Arch Dam, Utah.
*U.S. Bureau of Reclamation*

**Fig. 1.7** (*b*)
World Trade Center Towers, 110 stories, New York.
*Port of New York Authority*

**Fig 1.7** (c)
Jumbo jet hangar for American Airlines.
*Lev Zetlin Associates, Inc.*

**Fig. 1.7** (*d*)
Highway separation structures, San Francisco, Calif.
*PCA*

**Fig. 1.7** (*e*)
200,000-gallon Pedestal Sphere Water Tank, Westminster, S.C.
*Pittsburgh-Des Moines Steel Co.*

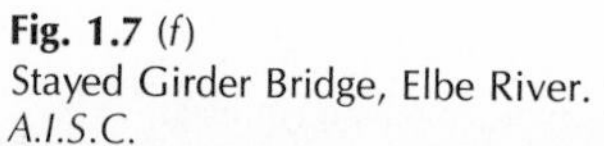

**Fig. 1.7** (*f*)
Stayed Girder Bridge, Elbe River.
*A.I.S.C.*

**Fig. 1.7** (*g*)
Static testing of F-14 aircraft.
*Grumman Aircraft Corporation*

**Fig. 1.7** (*h*)
210-ft. diameter space tracking and Telemetry antenna, Goldstone, Calif.
*Jet Propulsion Laboratory*

**Fig. 1.7** (*i*)
100 Constitution Plaza, 18 stories, Hartford, Conn.
*Bethlehem Steel*

**Fig. 1.7** (*j*)
Connecticut Yankee nuclear power station.
*Northeast Utilities Service Company*

**Fig. 1.7** (*k*)
Sewage treatment plant, Des Moines, Iowa.
*Environmental Protection Agency*

**Fig. 1.7** (*l*)
Space Simulation Chamber, Boeing Aircraft.
*Pittsburgh-Des Moines Steel Co.*

The structural engineer is also responsible for the creation of a structural system in harmony with each of the architectural alternatives. Approximate sizing of major elements in the structure is done to determine the feasibility and to estimate construction costs. In some buildings (e.g., a 10-story reinforced concrete frame and slab building) the engineer meets no great obstacles in evolving an economical solution. On the other hand, special buildings require a great deal of imaginative engineering. The 110-story World Trade Center towers in New York City (Figure 1.7 (*b*) have a structural system substantially different from that used in most smaller, high-rise buildings. The vertical elements of each tower are concentrated at the outer face of the building and in an inner core that contains the elevators and utilities. Beams span between the inner and outer vertical structures to form supports for the floor. The outer portion of the tower is made of closely spaced steel columns and is designed to act as a giant tube, cantilevered from the ground, in resisting the horizontal loadings produced by wind.

Another important class of structures consists of those built for industrial production and manufacturing. The design effort for industrial buildings is done by an engineering staff maintained by each company, or by a firm of consulting engineers. The design team will include many different types of engineers, with structural engineers having a major role. Pressure vessels, aircraft frames, spacecraft structures, and other special structures are usually designed by in-house structural engineers.

Civil engineering consulting firms hold the primary responsibility for certain types of very large and costly facilities such as nuclear power plants (Figures 1.7*j*). The engineering team is responsible for the total planning and design; again, the structural engineer is a key man in this team effort. He also becomes involved in environmental considerations, inasmuch as the design firm is asked to investigate and monitor large-scale environmental programs, including radiation protection and thermal pollution problems. The structural engineer faces a major responsibility with respect to safety. The determination of suitable design loads for a nuclear power plant presents many questions that are difficult to resolve. We obviously do not want release of radioactivity into the atmosphere, but is it realistic to design a power plant to survive the simultaneous occurrence of a maximum earthquake and the direct impact of a falling commercial plane? The design philosophy for this type of structure is necessarily extremely conservative, but one cannot overdesign to such an extent that the consumer cannot afford to pay for the power generated by the new power station. Finally, we note that nuclear reactors may contain unusual materials such as graphite, heavy aggregate concrete, and very high strength steels. The structural engineer must be familiar with their diverse properties and behavior under radiation and both normal and elevated temperatures.

Highway bridges and interchange structures are designed by state highway department bridge engineers and by structural engineering consulting firms. Our case study illustrates the latter arrangement in which the governmental engineers work with a private consultant. Structural engineers are also employed by governmental agencies to help plan and design a variety of structures other than bridges; included are locks and dams, flood control structures, and water supply and sewage treatment facilities.

Our earlier remarks have been directed to design of structures. A large number of structural engineers are engaged in other activities such as development and research. These activities include perfecting new components, materials, and structural systems from the idea stage to the point that they are acceptable for use in construction. Structural research and development groups and laboratories can be found in industry, in government, and in universities and colleges. They constitute a vital force in improving the structures of tomorrow.

### *Suggested Reading*

Cross, Hardy [1952]: *Engineers and Ivory Towers,* McGraw-Hill, New York.

Finch, James Kip [1960]: *The Story of Engineering,* Doubleday (Anchor Books), New York.

Gies, Joseph [1963]: *Bridges and Men,* Doubleday, New York.

Steinman, David B., and Watson, Sara R. [1957]: *Bridges and Their Builders,* Dover Publications, New York.

Straub, Hans [1964]: *A History of Civil Engineering,* MIT Press, Cambridge.

Talese, Gay [1964]: *The Bridge,* Harper and Row, New York.

## PROBLEMS

1.1 Find at least one small bridge in your community and investigate the structural system used. Does the bridge have hinged or roller supports or some other type of support system? Sketch a typical support detail and comment on how well it would satisfy the assumption of fixed, hinged, or roller support.

1.2 Sketch an idealized structural system for the classroom building in which you have your structures classes. Can you tell whether the building frame is made of structural steel, reinforced concrete, or timber? If it is structural steel, but the members are covered with concrete, why is the concrete there?

1.3 Find one old building and one building under construction on your campus. Compare the appearance of the structural systems and attempt to rationalize the difference.

1.4 Most field houses or gymnasiums on campuses have an exposed roof structure system that is clearly visible from the floor of the building. If you have such a building on your campus, study the structural system, sketching its elements and identifying the major force-resisting portions of the structure. Notice that the structure may have many secondary bracing members that are crucial in maintaining the integrity of the building. How does the building resist wind loadings?

1.5 List the structural and nonstructural features of the Solleks River Bridge project that are not typical for most bridges.

1.6 Obtain at least one sheet of actual engineering drawings (preferably from a local project) and study its contents. Sketch an idealized line drawing of the structure and comment on the types of supports for the main members (fixed, hinged, continuous, etc.).

## CHAPTER 2

# The Objectives of Structural Design

The Wheeling, West Virginia Suspension Bridge. One of the world's oldest existing major suspension spans, built by John A. Roebling in 1856. An ASCE National Historic Civil Engineering Landmark. *Wheeling News-Register*

# 2

## The Objectives Of Structural Design

A structure is created to serve a definite purpose. It may be required to enclose space, support a roadway or a machine, or contain or retain materials. It may travel to the moon and back, or it may be buried in the earth. If a structure is to fulfill its purpose, a number of design objectives relating to *safety, serviceability,* and *feasibility* must be specified and satisfied. In addition, we are usually concerned about the esthetics or looks of structures.

In order to meet these and other design objectives, we must have a basic understanding of how construction materials behave when stressed. Examples of common construction materials are rock and earth (foundations), concrete, steel, and timber. There are numerous others and the list is continually expanding. Section 2.1 contains a brief discussion of some of the most important characteristics of steel, concrete, and timber.

Among the most important of the design objectives is the provision of a safe structure. Collapse (as illustrated by the parking garage shown in Figure 2.1), severe distortion, buckling, or general yielding of a structural system is unacceptable under almost all circumstances, since such modes of failure may result in heavy economic losses and in the loss of human lives.

A number of approaches to the problem of structural safety are possible and are discussed briefly in Section 2.2. The most common of these, called *allowable stress design,* is based on the assumption that if the stresses under working loads are limited to values substantially smaller than stresses corresponding to failure, then safety is assured. Working loads are usually specified in codes of practice, but may be devised by the engineer. They are intended to represent the maximum load that will occur in the life of the structure and, when properly chosen, are rarely exceeded during the life of the structure. A common working load for classroom floors is about 60 pounds per square foot.

**Fig. 2.1**
Failure of a concrete parking garage.

Another approach to safety, one that is gaining in applicability, is termed *strength design, ultimate strength design, plastic design,* or *limit design.* These terms each have special meanings, but the methods of design have in common the assumption that behavior at failure or some other behavior threshold of either a cross section, a member, or the entire structure can be understood sufficiently well to be predictable. The designer ensures that there is a sufficient margin of safety between the expected working loads and the calculated failure loads. He first arrives at the required failure capacity of the structure by increasing the working loads with load factors greater than unity, and then designs the structure to ensure that this required capacity is exceeded.

Some structural systems require the use of both the allowable stress and the strength design methods because knowledge of stresses may be no guarantee of a suitable margin of safety against failure, and the prediction of the failure load may not guarantee satisfactory performance under working loads. The design of prestressed concrete structures and of aircraft and space structures are examples of such situations.

In addition to safety against failure, a structure must have adequate serviceability. This means that all aspects of performance must be acceptable for the intended use. Deflection and cracking must be limited so that they are virtually invisible to the layman. Vibration and noise should be controlled. Liquid and gas containers must not leak, and foundations must not settle excessively. Serviceability requirements constitute an almost limitless list and must be carefully tailored to the needs of the particular structure. The key to the satisfaction of the serviceability requirements is a full understanding of the behavior of the structure throughout all of its loading and environmental phases.

A structure may be safe and have adequate serviceability, but all of this is useless unless the structure is *feasible.* Economic feasibility is of paramount importance. Once this is established, the structure must be capable of being constructed. Many design decisions are dependent on construction methods, and the structural engineer must be familiar with these.

The structural engineer is responsible to a client who may not be expected to have the necessary skills and knowledge to fully understand the principles of structural design. The engineer is also responsible to society as a whole, since his decisions affect their safety and the quality of their environment. Under these conditions, *caveat emptor* of the marketplace is inappropriate. Instead, there are professional and legal controls on the decision powers of the designer. In the case of common types of structures, the profession creates design codes of practice that reflect current viewpoints on strength, serviceability, and construction practices. These design codes are often enacted into law by the various political jurisdictions, thus providing a legal means of protection for the client and the general public. Unusual structures such as nuclear containment vessels, tunnels, special bridges, and the like are not usually covered by such codes. In these cases the reputation and professional responsibility of the engineer are the means of protection.

Codes of practice vary widely in style, objectives, and approach. Some contain detailed requirements regarding load, stresses, strength, design formulas, and other details. Others, called *performance codes,* specify the required performance of the completed structure but do not attempt to prescribe the method of achieving that performance. Such codes are gaining importance and will play a large role in future design. They assign more responsibility to the engineer, and should elicit more imaginative designs.

## 2.1 MATERIALS

A thorough understanding of materials is basic to understanding the behavior of members and structures. Materials are discussed in more detail in Volume 3, but a brief discussion of the important physical properties of common materials is appropriate early in a study of structural engineering. Although most structures are subjected to complex, two- and even three-dimensional stress states, the basic structural properties of building materials are determined on simple specimens loaded in either tension or compression.

Steel is used for a diverse assortment of structures. It has nearly identical properties in tension and in compression; thus its most important properties can be determined from tension tests on small specimens. Figure 2.2 indicates a stress-strain curve for a typical structural carbon steel. This particular steel has a yield strength of 36 ksi, but steels are available for a wide range of strengths up to around 100 ksi and more. Note the definitions of the various stages of stress and strain in Figure 2.2. The slope of the stress-strain curve is a measure of stiffness; it is called the modulus of elasticity, $E$, and has a value of about $29 \times 10^3$ ksi for all steels. $\epsilon_y$ is the strain at yield, and $\epsilon_p$ is the plastic strain.

One of the most important characteristics of carbon steel is its duc-

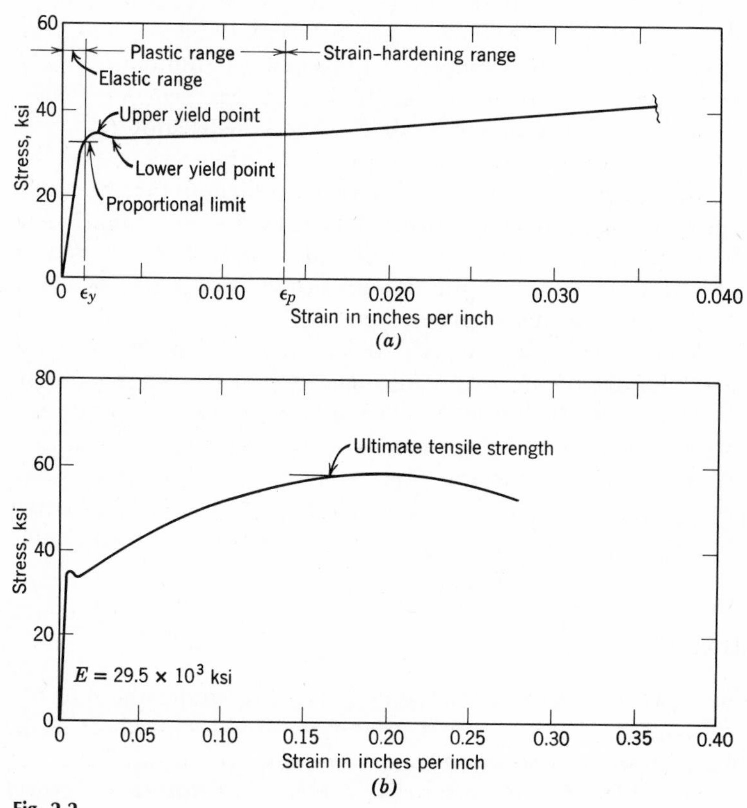


**Fig. 2.2**
Stress-strain curve for typical A-36 carbon steel. (*a*) Strain range 0 to 0.040 in./in. (*b*) Strain range 0 to 0.40 in./in.

tility. After yielding, the steel specimen can strain a large amount (on the order of 20%) before final rupture. The ultimate tensile strength is about 50% greater than the yield strength for the particular stress-strain curve shown here. Steels differ in this regard, and their individual characteristics affect the choice of safety levels. Even ductile steels may fail in a brittle manner under certain conditions, such as at low temperatures.

Concrete is a completely different material than steel, and it is not surprising that methods of design differ from those for steel. Concrete is very weak in tension in comparison to its compressive strength. Figure 2.3 is a typical stress-strain curve based on compression tests on cylinders. Practical compressive strength, $f_c'$, varies from 3 to 7 ksi, but higher strength mixes can be made. The modulus of elasticity for concrete increases with concrete strength and has a value about one-tenth that of steel. Concrete does not have a definite yield strength, but rather the stress-strain curve continues smoothly to the point of rupture at a strain of about 0.004 in/in. The lack of linearity of the stress-strain curve at higher stresses poses some problems in analysis because the resulting behavior of concrete members is nonlinear. Section 5.1 contains a discussion of linearity and nonlinearity. Usually we can assume linearity in the analysis of reinforced concrete structures.

Because of the low tensile strength of concrete, we place steel reinforcing bars in regions of tension in a concrete member. The cross section then acts as a composite one with the two materials, steel and concrete, working together. The study of reinforced concrete in Volume 3 of this text involves an understanding of this composite behavior.

The behavior of wood is unique and is dependent on the grain and cell structure of wood. Small clear specimens of wood are tested to determine basic wood properties. The behavior is linear in the useful stress

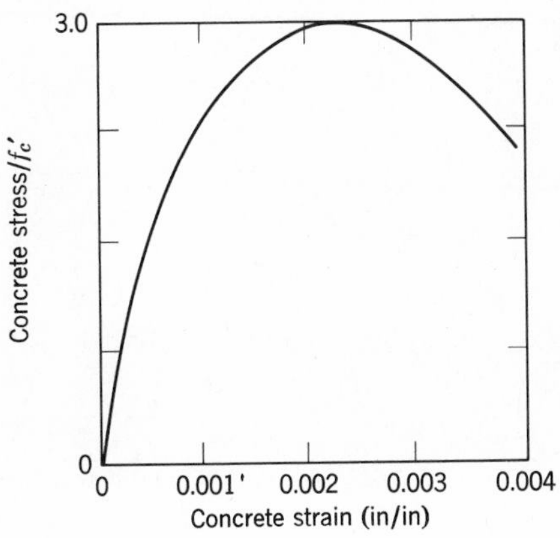


**Fig. 2.3**
Compressive stress strain curve for concrete.

range. Wood is not ductile like steel, but it does have a high capacity to absorb energy resulting from rapidly applied loads. The modulus of elasticity of wood varies with the species; a typical value is $1.8 \times 10^3$ ksi.

Wood has the advantage of high strength-to-weight ratio, so that its use in long span structures is advantageous. It is easily worked, and because of modern gluing technology wood members may be fabricated into unlimited sizes and shapes. Wood does not lose its strength when heated, and large wood members are often more fire resistant than the equivalent steel sections.

Problems associated with its grain structure must be accounted for when designing in timber. Shear strength along the grain is low, as is tensile strength across the grain. A full understanding of the advantages and weaknesses of wood permits the design of many beautiful and durable timber structures, such as the one shown in Figure 2.4.

Although there are many aspects peculiar to steel, concrete, or timber behavior, it is interesting to note that in steel design the buckling of members or thin plate elements is a primary problem; in concrete design, cracking, creep, dead weight, and ductility are of concern; and in the case of timber, the design of connections is of great importance.

**Fig. 2.4**
Laminated timber skating rink. *Unadilla Laminated Products*

## 2.2 SAFETY

Requirements of safety represent the principal responsibilities of the structural engineer because they are the most direct means of protection of life and property. To fulfill this heavy responsibility, the engineer must fully understand the environment and use of the structure and the behavior of the materials of construction. He then uses this knowledge to predict structural behavior. An important part of this process is the judgment needed to make appropriate allowances for the uncertainty involved in the prediction of load effects and structural response.

When considering safety requirements it is useful for the engineer to envision a number of successive load stages to be imposed on the structure, each stage being more severe than the last. If performance of the structure through each load stage is considered in detail, the engineer can truly achieve understanding of his structure.

The first load stage is that corresponding to loads which are expected to occur repeatedly. These may be actual floor loadings, normal storm wind loads, or truck loads of legal magnitudes. They may be aircraft or spacecraft loadings under conditions of normal maneuvers. Such loads occur frequently in the life of the structure, and under them the structure is required to perform indefinitely without distress. Virtually all of the many structures that you see every day are operating under these conditions.

A second load stage consists of large loads that might occur only a few times in the life of the structure. These loads often correspond to working loads specified in design codes. The loaded structure is still required to perform adequately with no signs of distress; the engineer is profession-

**Fig. 2.5**
Load test of a pedestrian overpass.

ally and legally responsible for satisfactory performance of the structure. Figure 2.5 shows a pedestrian footbridge under a simulated design working live load. The span is loaded with pallets of cement in the load test shown here. Simultaneous occurrence of several types of such loadings is very unlikely, and for this reason the required safety margin against collapse may be reduced under such conditions. An example would be the simultaneous occurrence of the full design working live load and full design wind load.

The third stage consists of severe and rare loads that might possibly act during the life of the structure but for which it is not reasonable to expect flawless structural performance. Included in this category are extreme wind loads such as tornados, rare extra-legal truck loads (which are occasionally found on the highways in spite of the existence of weigh stations), and loads resulting from moderately large earthquakes. A well-designed structure is expected to carry such loads without loss of life and with minor repairable damage. Figure 2.6 indicates a prestressed concrete beam loaded beyond its "cracking load." Notice the excessive deflection and the cracks near midspan (shown marked by heavy black lines). The beam is still capable of carrying increased load. Some damage in the third load stage is acceptable because the economic consequences of the damage may be less than the cost of providing a stronger structure to avoid the damage. Major problems exist in attempting to find the proper balance between cost and strength under these circumstances. In this load region, the responsibilities of the engineer are shared with those responsible for control of loads.

The last loading stage is that corresponding to imminent failure. This

**Fig. 2.6**
Prestressed beam.

*E. G. Nawy and J. G. Potyondy*

**Fig. 2.7**
Short steel column test.

*Bethlehem Steel*

is a hypothetical stage in real structures but it is one that must be considered if behavior at failure is to be understood. In order to study behavior in this load range, tests are conducted. Figure 2.7 indicates a test of a short steel column loaded beyond the yield level and Figure 2.8 indicates a long steel column buckling laterally in its failure mode. In this case, stresses are less than yield, but failure is imminent. The engineer must be able to control and predict the mode of failure. His ability to accomplish this has a profound effect on the required safety levels, since a properly directed gradual failure mode may ensure that human occupants can be evacuated before collapse. The consequences of failure are therefore much reduced, permitting a more economical approach to design.

Another motivation for detailed understanding of failure processes is that a large amount of energy can be absorbed by a material as it is stressed inelastically beyond the proportional limit. This means that a substantial part of the total energy of an earthquake or other action on the structure is absorbed in the last stages of structural deformation before collapse. Thus, with adequate ductility in both members and connections, structural collapse may be avoided, permitting evacuation and possibly repair.

Failure may occur in any of a number of ways and all of these must be

**Fig. 2.8**
Long steel column test.

*Bethlehem Steel*

understood by the designer. A common type of failure is that of yielding of the material. This mode is accompanied by large deformations and usually some redistribution of forces within the structure. Another type of failure is instability or buckling. This may occur at very low stresses, and may or may not be accompanied by visible deformation. Some structures, such as those made of thin plates, may undergo buckling but still have substantial capacity even after buckling has occurred. This behavior is called postbuckling action; it is often utilized in design. Failure may also occur by brittle fracture resulting from various causes, such as rapid application of load, shape effects causing stress concentrations, low temperature, embrittlement by chemical action, or stress corrosion. Effects of creep may cause failure by inducing undesirable deformation in the structure. Creep is a time-dependent phenomenon where the strain in a material increases under constant loading; it is a definite problem in concrete structures. Finally, the effect of stress reversal or repetitive applications of stress through many thousands of cycles can be a source of failure. This phenomenon, known as fatigue, must be considered in the design of aircraft and spacecraft, highway and railway bridges, crane runways, and other structures that serve a load mission repeatedly.

Although structural performance may be considered through all the load stages discussed, code requirements often focus on certain key stages, with the expectation that satisfaction of code criteria at one or two stages will ensure acceptable performance at all stages. Two viewpoints are commonly found in specifications and codes: the first of these is based on an allowable stress philosophy; the second on a strength philosophy.

### Allowable Stress Design

Allowable stress design begins with the choice of appropriate design working loads. The nature and magnitude of design working loads are based on the type of structure. For example, the American National Standards Institute (ANSI), through its committee A58, sets design working loads for buildings. These are only guides and have no legal significance. The Basic Building Code (BOCA[1970]) adopts the loadings set by the ANSI as design working loads. The Basic Building Code is then enacted into law by various cities and states. When legally adopted in this manner, the loads become legal minimums, and the structural engineer is required to use them as such in his design. Other examples of design working loads are those specified by the American Association of State Highway Officials. This association issues a design code specifying highway bridge loadings and design methods for highway structures. The states may then adopt the loadings, perhaps with modifications, as standards for structures in their jurisdiction. Codes may not cover all possible situations. The Solleks River Bridge, discussed in Chapter 1, was designed for heavy logging trucks. The truck loading in this case was agreed on by the engineer and his client, and was chosen to represent the heaviest likely logging truck. Loads are discussed in detail in Chapter 3.

Once design working loads are established, the structure is analyzed under assumptions of elastic behavior and stresses are determined. Allowable maximum stresses are specified by appropriate code-writing bodies. Such allowable stresses are chosen as a fraction (about 60%) of the stress corresponding to failure by whatever failure mode governs. This may be the buckling stress, the fatigue limit stress, the yield stress, or another stress associated with failure. This procedure ensures that in normal service the structure will behave nearly elastically and will not be subject to buckling, fatigue, or other failure modes.

Design of aircraft and spacecraft structures follows an allowable stress philosophy. The design working loads are accurately determined for the maximum conditions encountered on the mission. Allowable stresses under these loads and their possible combinations are set very close or equal to the yield stresses, or to stresses corresponding to other failure modes. This is justified by careful and accurate testing and control of materials. In addition to this procedure, a strength approach is used requiring that ultimate strength be sufficient to resist a loading about 1½ times the critical design working load.

### Strength Design

Strength design begins with a determination of the design working loads, just as in allowable stress design. These loads are then multiplied by *load factors* to determine the required strength. As we have mentioned, the load factor for aircraft and spacecraft structures is about 1½. The factor for civil engineering structures is in the same range, but may vary in accordance with the type of load or load combinations.

Structural steel design for certain types of buildings and bridges can be carried out using a strength approach. This is called *plastic design* because of the plastification that occurs in the steel cross section when applied moments are equal to the flexural yield strength of the section. Concrete design by this method is called *strength design.* For concrete structures, the inherent variability of materials and field dimensions necessitates the use of a *capacity reduction factor* ranging from about 0.65 up to 0.9; the factor is applied to the theoretical member strength to estimate the minimum strength actually provided.

### Future Approaches to Structural Safety

All of the methods of structural safety are based on the fact that a margin of safety is needed to cover uncertainty in loading, materials, methods of analysis, and quality of construction. The theory of probability may be used to model problems of uncertainty, and shows promise of contributing to the solution of problems of uncertainty in structural safety in the near future. Some progress has already been made, and methods are now available that may assist in the determination of load factors or other parameters that occur in safety decisions. Structural engineering is strongly dependent on intuition and judgment based on experience, so that safety decisions cannot be made solely with analytical models of probability and mechanics. The decisions of the experienced engineer will always remain the most important factor in design.

A discussion of means of achieving safety is hardly complete without some consideration of the consequences of failure. A knowledge of some catastrophic failures of the past may be useful to the engineer who must design for the future.

**Fig. 2.9**
Failure of the Second Narrows Bridge, Vancouver, B.C.
*F. M. Masters*

One of the more serious structural failures in recent times was the collapse during construction of the Second Narrows Bridge, Vancouver, B.C., in 1958. Figure 2.9 shows the structure during construction and after failure, which occurred suddenly and resulted in the deaths of nearly 20 men. Several design errors in the temporary supporting grillage shown in the lower right photograph were responsible for the failure.

Another bridge failure, with a loss of 75 lives, took place when the Quebec Bridge collapsed during construction on August 29, 1908. This failure is discussed in Section 5.1 and illustrated in Figure 5.4.

The Second Narrows and Quebec bridges failed during construction. The recent tragic failure of the U.S. 35 Highway Bridge at Point Pleasant, W. Va., occurred in December 1967 while the structure was loaded with vehicles. The failure was caused by brittle fracture of a critical eyebar component, and is discussed in Chapter 19, Volume 3.

Although bridges have had some spectacular failures, especially during construction, all types of structures are vulnerable. Dams have had a particularly high mortality rate. In the first 60 years of this century, about 1650 dams over 45 feet high were constructed in the United States, and 1.8% of these failed with a release of water. Over 400 deaths were caused by these failures. There have been major dam failures in other countries. For example, the Malpasset Dam in France failed in December 1959 with a loss of about 400 lives (see Figure 2.10). Dams are particularly vulnerable

**Fig. 2.10**
Malpasset Dam failure.

*Civil Engineering, ASCE*

to the uncertainties associated with foundations and earthwork. Dam safety has been the subject of much discussion in recent years (see Biswas and Chatterjee [1971], and Golzé [1971]).

Buildings have also had their share of failures. Lack of adequate bracing of steel frames and shoring of concrete have been common causes. Although most failures occur during construction, some do happen while the building is in use (see Figure 2.1).

## 2.3 SERVICEABILITY

Many of us have felt the annoying effects of motion or vibration in a building floor or other structure as produced by an escalator, machinery, or simply by people moving about. We have also observed that heavily loaded beams deflect downwards, often with an accompanying unsatisfactory appearance. Such obvious signs of poor structural performance are uncommon. Usually a structure appears to be rigid and unmoving even though every load-carrying element is subjected to stresses and corresponding strains. The cumulative effect of strain is displacement of the structure from its original unstressed position. A major activity of the structural engineer is to understand, control, and direct this important behavior to obtain a structure that performs in an acceptable manner. The trend to stronger materials, with correspondingly lighter and thinner sections, brings with it increasing problems of vibration and deflection.

The most obvious manifestation of strain in a structure is the displacement under static load. Under given loads this quantity can easily be calculated, and is often limited as a basic structural requirement. A downward sag in a beam is unsightly and lends an appearance of weakness to the structure even though the strength may be quite adequate. Displacement limits may be imposed, then, to ensure satisfactory appearance. Beams are usually cambered, or built with an initial curvature opposite to that caused by loads, so that displacements under load will not appear large. Figure 2.11 shows a continuous plate girder bridge cambered upwards to counteract the effects of dead load (concrete deck) still to be added.

The "feel" of a floor may also be controlled by limitations on static displacement. A floor with perceptible bounce may be considered inadequate by the users. This aspect of behavior is closely related to the visible deflection, since neither actually indicates weakness, but both give a psychological feeling of weakness.

Another important controlling factor for deflection requirements is related to the interaction of structural components with architectural items. A glass window under a spandrel beam would crack if even the slightest beam deflection were imposed on the window edge. For this rea-

**Fig. 2.11**
Spences Bridge, B.C.: a continuous steel girder.

son, windows are mounted in a flexible edge support that can accommodate a certain amount of deflection without loading the window. The beam deflection is then limited to suit the window detail. A similar effect is found when plaster ceilings are applied under beam spans. Experience has shown that the plaster, which is stressed in tension as the beam deflects, will crack when the deflection exceeds about 1/360th of the span. Because of the extensive use of plaster ceilings in the past, the fraction 1/360 has become a common specification for the live load deflection of floor and roof beams.

Deflection may be limited for reasons other than appearance or compatibility with building materials. An example is the interesting effect known as *ponding* on flat roofs, discussed in Section 3.5. Displacement becomes a critical part of the design for many types of antenna structures. Microwave towers support parabolic reflectors that are aimed at a source signal which may be many miles away. Angular deviations from the correct aim must be carefully limited to avoid loss of the signal. Radio telescope structures, aimed to receive signals from space, are sensitive to displacements for similar reasons. Many such structures are movable, which further compounds the problem. The shape of the reflecting surface must be maintained to close tolerances to pick up the weak signals from faraway sources in space.

Displacement is not always difficult to see. Modern airplane wings undergo visible displacements under normal operating conditions. When you travel by air, you might find it interesting to sit in a position from which you can observe the action of the wings during the trip.

The determination of static displacement and its prevention by requirements for adequate stiffness is perhaps one of the simpler problems for the structural engineer. Problems of vibration, however, are likely to be difficult and unpredictable. The vibration resulting from rotating or reciprocating machinery can be harmful to the structure in addition to being unacceptable from the viewpoint of human comfort. The natural frequencies of the structure, which depend on the mass and the stiffness, must be controlled in order to avoid destructive vibration. Wind and earthquakes can cause further difficulties. Here the loading is random and the response of the structure is a complicated random process. Vibration, although discussed here under the topic of serviceability, may be destructive.

Although more a problem of strength than of serviceability, a famous case of excessive structural vibration is that of the first Tacoma Narrows Bridge. This structure, with a main span of 2800 feet, was completed in July 1940, at a cost of $6.8 million. At the time, it was the third longest suspension span in the world. The slenderness of the suspended deck represented a distinct departure from earlier suspension bridge designs. The bridge had shown vibratory tendencies even during construction, and had earned the name "Galloping Gertie" among motorists who used the span. On November 7, 1940, under the action of a 42-mph wind, the structure began to oscillate with growing magnitude and in several modes of vibration. Professor F. B. Farquharson of the University of Washington was at the site, and the following report of the tragedy is based in part on his eyewitness account.

> ... this wind caused a vertical wave motion that developed a lag or phase difference between opposite sides of the bridge, giving the deck a cumulative racking or side-to-side rolling motion.
>
> Failure appeared to begin at mid-span with buckling of the stiffening girders, although lateral bracing may have gone first. Suspenders snapped and their ends jerked high in the air above the main cables, while sections of the floor system several hundred feet in length fell out successively, breaking up the roadway to the towers until only stubs remained (*Engineering News Record* [1940]).

Figure 2.12 shows the bridge in a state of collapse.

The United States Federal Works Agency, which had partially financed the bridge, appointed a Board of Engineers consisting of Amman, von Kármán, and Woodruff to report on the failure. A partial summary of their conclusions (which differs slightly from the eyewitness report) follows:

> The immediate cause of the failure of the Tacoma Narrows Bridge was the slipping of the center cable band on the north cable. This slippage initiated a torsional vibration of the bridge structure. Design of the bridge was such that torsional vibration, once established, tended to increase. This torsional movement caused bending stresses in the concrete floor, stressed the structural members beyond their elastic limits, and created impact loads on the suspender cables under which one of them broke; a progressive snapping of the suspender cables then followed. (*Engineering News Record* [1941]).

**Fig. 2.12**
Failure of the first Tacoma Narrows Bridge, Tacoma, Wash.

*Wide World Photos*

The Tacoma disaster provided a great impetus to research in the field of aerodynamic stability and structural vibration. This work led to modifications on the Golden Gate Bridge and several other large bridges of the suspension type, and to a far better understanding of structural behavior. However, problems of vibration resulting from wind still come back to plague the designer. There are many examples of modern structures that have required modifications after completion because of unpredicted vibrational response due to wind. One recent case was that of the Severn Suspension Bridge in England. The unique triangulated suspenders vibrated excessively, and it was found necessary to install a damping system. One of the objectives of all designers must be to understand and to avoid behavior such as wind-induced vibration.

Serviceability includes a number of additional aspects of performance. The structure must be durable. Materials must be chosen with regard to effects of corrosion or deterioration. All materials have their advantages and disadvantages in this regard. Steel is subject to corrosion but may be painted or galvanized. Some modern steels, called "weathering steels," form a tough oxide that resists corrosion. Timber may rot or delaminate. The engineer must be familiar with wood preservation technology and with design techniques to prevent delamination. Concrete may deteriorate as a result of poor aggregates, poor cement, or chemical reactions set up by roadway salt or other causes. The objectives of the designer must include a careful control of materials, based on an understanding of their performance in the appropriate environment.

## 2.4 FEASIBILITY

The construction of a proposed structure must be economical as well as possible. A basic rule in design is that the designer should have in mind at least one method of construction. The structure may not be built in exactly the manner anticipated, but still the feasibility of the structure will have been supported. The designer who is familiar with construction problems and methods will facilitate economical construction without imposing undue control over the construction process.

The Solleks River Bridge, discussed in Chapter 1, was made of precast elements in order to avoid transportation of concrete over the long distance to the nearest source of supply. The maximum length of a single element, about 78 feet, was chosen to allow vehicles to navigate the narrow, twisting access road to the site. The weight of the largest elements was within allowable loads for highway bridges, and the size allowed passage under highway overpasses. These and other details were carefully chosen to facilitate construction and to meet the objective of feasibility.

## 2.5 DESIGN CODES

Structural requirements are partly set by design codes that are written for the broad protection of society as a whole as well as for the owner of a single structure.

The first available code governing the practice of construction was a part of the Code of Hammurabi, a King of Babylon, and was written about the year 2040 B.C. This code derived its authority from the fact that it was written under the direction of the God Marduk. Figure 2.13 shows a stone monument, now in the Louvre, Paris, depicting the God of Justice, Shamash, handing a tablet of the Code to Hammurabi, who is standing on the left.

Hammurabi's Code established strict standards of performance, as may be inferred by its provisions regarding building construction (Driver and Miles [1955]):

> If a builder has made a house for a man and has completed it for him, he shall give him 2 shekels of silver for every SAR of the house for his fee.*
>
> If a builder has made a house for a man and has not made his work sound,

* This is equivalent to 72 days pay for 53 square yards of house, a figure not unlike today's housing prices.

**Fig. 2.13**
The Code of Hammurabi.

*Publications Filḿees D'Art & D'Histoire*

and the house which he had built has fallen down and so caused the death of the householder, that builder shall be put to death. If it causes the death of the householder's son, they shall put that builder's son to death.

If it causes the death of the householder's slave, he shall give slave for slave to the householder. If it destroys property, he shall replace anything that it has destroyed; and, because he has not made sound the house which he has built and it has fallen down, he shall build the house which has fallen down from his own property.

If a builder has made a house for a man and does not make his work perfect and the wall bulges, that builder shall put that wall into sound condition from his own silver.

Design codes have been with us ever since the time of Hammurabi. They provide a framework within which the individual engineer makes detailed design decisions. At the same time, they represent a valuable collective opinion of the profession as a whole in meeting its responsibilities to society. Codes operate either by providing for minimum requirements on structural details and on methods of analysis and design, or by specifying the performance of the finished product. It is interesting to note that Hammurabi's code was a *performance code,* in that it was only concerned with results, and had no mention of details or methods. The structural engineer uses a number of design codes in his practice. Some of the more widely used codes and specifications, including those of the American Concrete Institute and the American Institute of Steel Construction, are shown in Figure 2.14.

**Fig. 2.14**
Typical design codes.

Structural design involves many objectives; one way of summarizing them is to repeat a statement made by Professor Jack Benjamin of Stanford University at a recent meeting on structural safety. In response to a comment regarding the professional responsibilities of the engineer, he stated, "The client buys the structure for a few years, and may sell it at any time. The structural engineer buys it and all its problems for his personal lifetime."

## PROBLEMS

The following problems do not necessarily have "correct" answers, but are posed to motivate discussion and thought on some aspects of structural engineering discussed in this chapter.

2.1 Discuss the similarities and differences in structural safety approaches for spacecraft and for steel buildings.

2.2 Investigate the principles of prestressed concrete behavior in sufficient depth to discuss the reasons for the necessity of using both allowable stress and strength methods.

2.3 Does the concept of probability of failure fit your present understanding of structural performance?

2.4 Give three examples of problems of structural serviceability which are from your own experience or observations.

2.5 Design and build a model that demonstrates the effect of ponding.

2.6 Hammurabi's code was a performance code. Discuss other codes with this classification in mind.

2.7 Are legal design codes really necessary?

2.8 Many structural failures are reported in the weekly magazine *Engineering News—Record*. Study any four consecutive issues of the magazine and make a list of the reported failures and their probable causes. Can you draw any conclusions about common reasons for failure from your limited study?

2.9 Why are two scales needed to depict the stress-strain behavior of carbon steel as in Figure 2.2?

2.10 A skyscraper deflects horizontally under wind loadings. What amount of lateral deflection, or drift, would you consider acceptable at the top of a 500-foot office building? If you worked in this building every day, would your answer be the same? Is the maximum drift the important quantity, or is it the maximum velocity or acceleration?

2.11 The lives of tens of millions of people are dependent upon the safety of the support cables in elevator systems. What factor of safety would you recommend for designing elevator cables? After thinking about this problem, consult your library or local building official for the values actually used, and discuss the answers.

2.12 Pick any three structures at random from your locality and inspect them visually for any signs of distress or lack of serviceability. Report your findings and discuss possible reasons for any shortcomings that you observed.

2.13 List the factors that should influence the selection of safety factor. Give three types of structures for which you would recommend relatively large safety factors and three for which relatively small factors would suffice.

2.14 A legal building code exists in your city. Who is responsible for this code? Does it apply to all structures built in the city? Does it refer to or use any of the codes shown in Figure 2.14?

CHAPTER 3

# Loads: Man-made and Natural

*British Columbia Forest Service*

# 3

## Loads: Man-made and Natural

The detailed design of structures consists of the determination of configurations and sizes of members and connections, and is primarily governed by the requirement that structures must safely carry all loads imposed on them. Therefore, a knowledge of the expected maximum loads and load combinations is essential to the design process. The discussion presented in this chapter is not intended to provide an exhaustive reference on loads, but instead will illustrate the concepts of load analysis.

Structural engineers must determine the rational combinations of loads that will produce maximum stresses or deflections in various parts of the structure. It is not feasible to design ordinary structures to resist all conceivable combinations of loads and exceptionally large forces (such as tornados and large earthquakes); thus the design is, by necessity, uncertain. Statistical and probabilistic evaluation of load intensities and structural performance with adequate accounting for financial losses and human injury may be possible, but at present these factors are only beginning to be considered quantitatively by structural engineers. Instead, the magnitude of the loads and their critical combinations are established by judgment based on experience, measurements, and logic.

Although the load values given in codes may be legal minimums, they often serve only as guides. In some cases even such general guides may be insufficient; for example, the accumulation of snow in certain geographic locations or on novel curved roofs is highly variable. The wind pressure on tall structures depends on a great number of factors that cannot be described in general terms. New types of structures such as monorail supports and aerospace installations are subjected to unique loadings that are not covered in any code. Thus we see that the determination of the critical or controlling loads is often not a routine matter and may require extensive study and even actual testing. The specifications for large projects frequently contain loadings that differ from the applicable city or state building codes. In the case of unique projects the design engineer develops criteria for loads which may be submitted for review to an advisory group of consulting engineers.

To simplify the design of common structures, building codes specify minimum design loads for the various uses of a structure. The magnitude of *design loads* (i.e., the loads to be used in determining the sizes of structural members) has been established by many years of practice and, to a lesser extent, from research investigations. What load values would you specify in codes? If the load is a maximum possible value (e.g., the heaviest truck-trailers closely spaced along the entire bridge), the cost of construction would be very high. It is obvious that probabilistic considerations must enter, directly or intuitively, into the determination of design loads. Yet, the design values of live loads are very rarely reached during the lives of most structures since the specified loads are based on conservative estimates. Even full crowding of a balcony by an all-star football team would not greatly exceed the value specified by codes (compare your estimate of the weight of packed people to the design load of about 100 psf).

In some instances the full design load does occur often, for example, the active earth pressure against retaining walls, truck loads on short bridges, and the water load on various hydraulic structures.

## 3.1 LOAD TYPES

Loads are usually classified into two broad groups: *dead loads* and *live loads*. Dead loads (DL) are essentially constant during the life of the structure and normally consist of the weight of the structural elements. On the other hand, live loads (LL) usually vary greatly. The weight of occupants, snow, and vehicles, and the forces induced by wind or earthquakes are examples of live loads. The magnitudes of these loads are not known with great accuracy and the design values must depend on the intended use of the structure.

A deeper appreciation of the basis for load analysis can be gained by considering a different load classification: those loads that are man-made and subject to some degree of control, and those that result from natural phenomena and therefore cannot be controlled. Man-made loads can be regulated, although we may choose not to do this, whereas our approach to the analysis of natural loads must be limited to understanding and prediction based on probabilistic studies.

For the purposes of structural analysis, we can idealize any load as one of three kinds: (a) *Concentrated loads* that are single forces acting over a relatively small area; for example vehicle wheel loads, heavy machinery, column loads, or the force exerted by a beam on another perpendicular beam. (b) *Line loads* that act along a line. The loads exerted by a train may be idealized as a line load, in units of force per unit length. The weight of a partition resting on a floor slab is also a line load. (c) *Distributed (or surface) loads* that act over a surface area. Most loads are distributed or are treated as such, for example, wind or soil pressure, and the weight of floors and roofing materials.

The principal types of loads will be discussed in the following categories:

1. Dead load
2. Occupancy loads (persons, machines and supplies)
3. Bridge live loads
4. Snow and rain
5. Wind and blast
6. Earthquake
7. Water and earth pressure, wave and ice
8. Loading induced by temperature, shrinkage, and lack of fit

We shall discuss the nature of each of these loading types in the remainder of this chapter, indicating examples of actual design values and also tracing, to some extent, the development of commonly used code provisions.

## 3.2 DEAD LOADS

Dead loads include the weight of all permanent components of the structure, such as beams, columns, floor slabs, roofing, and bridge decking. They also include architectural components, such as ceiling tile, window fixtures, and room partitions. Fixed furniture or equipment is usually given a separate classification, although its effects are the same as those of dead load.

Dead load is perhaps the simplest of all loading types to handle, since it can be readily computed from given dimensions and known material densities. However, the structural dimensions are not known during the initial design phases, and estimates must be made which may be subject to later changes as the structural proportions develop. With some experience, and with approximate analysis techniques of the type discussed in Chapter 7, one can quickly estimate a preliminary design from which reasonable dead loads can be computed. In other than routine cases, this approach may have to be supplemented by reevaluation of the dead load and a redesign of the structure.

The unit weights of some of the more common building materials are indicated in Table 3.1. Some of these are average values and large variations are possible. Dead loads are computed on a conservative basis, which often involves the use of estimates slightly on the high side so that redesign is unnecessary for minor changes in the final proportions. This

**Table 3.1 Unit Weight of Construction Materials**

| Per Volume | | Per Surface Area | |
|---|---|---|---|
| Material | $Lb/Ft^3$ | Material | $Lb/Ft^2$ |
| Aluminum | 165 | Insulation | 2 |
| Asphalt | 80 | Roofing | 5-10 |
| Brick | 120 | Tile | 10-15 |
| Clay | 80 | Brick partitions (4 in.) | 40 |
| Concrete | | Gypsum block partitions (6 in.) | 18½ |
| light-weight | 90-110 | Clay tile partitions (6 in.) | 28 |
| semi-lightweight | 120-130 | Hollow concrete block walls (6 in.) | 43 |
| normal | 150 | | |
| Sand and gravel | 90-105 | | |
| Steel | 490 | | |
| Timber | 35-40 | | |

(An extensive set of material weights is given in AISC [1970])

procedure must be used with care, however, for there are many cases in which the dead load assists the structure by counteracting the live load. For example, the tension in some of the legs of a tower and the overturning action resulting from wind loads are usually reduced by the dead load. The maximum tension is found by using a dead load value in analysis that is at least as low as the actual dead load. This situation also occurs in building frames, especially during construction before the full dead load is acting.

Some DL stresses may reach their maximum values when the structure is only partially completed. The analysis of forces during erection is especially significant in the case of bridges. An excellent example can be seen in Figure 3.1, where "cantilever construction" of a bridge is shown. Similarly, DL effects may become critical during demolition or alteration work because the structure is taken down in an improper order; for example, if bracing (walls, diagonal bars) is removed too soon. A relatively large number of collapses have occurred during demolition of structures. For this reason, more and more local codes require the approval of demolition procedures by a licensed professional engineer.

## 3.3 OCCUPANCY LOADS

Occupancy loads include all loads that are directly caused by humans, machines, or movable objects. These loads usually act only during a fraction of the life of the structure, yet it is necessary to design for conservatively high values.

**Fig. 3.1**
Cantilever construction: the Poplar Street Bridge, St. Louis.

*Bethlehem Steel*

Occupancy loading defies any truly rational description. To aid the designer, specifications usually prescribe uniformly distributed live loads that conservatively represent load maximums. Table 3.2 shows an example of a specification of uniform live loads in buildings, according to the American National Standards Institute Code (ANSI [1955]). Most other codes give similar values, although wide variations are found. To simplify the analysis and because of the lack of any better information, live loads are spread over the entire floor area as uniform loads, even though the actual loads may be concentrated in a localized area.

The occupancy loads tend to be conservative. For example, assuming that a typical person weighing 170 lb occupies an area of 2.25 ft$^2$ (the area of four 9-in. floor tiles), a live load of only about 75 psf is obtained. It is possible to crowd people to produce a load of as much as 125 psf, but obviously such "full" occupancy over a large area is most unlikely. Failure of floors due to overloading is exceptionally rare. The live loads in Table 3.2 include such minor items as the weight of furniture. As a comparison, the DL on a floor of a typical building is about 85 psf (75 for a 6-in. concrete slab, 5 for flooring, and 5 for the ceiling below).

In warehouses, the LL magnitude depends directly on the use of the structure; hence it is advisable to post the design live loads in order to prevent a new tenant from overloading the floor. Since designs are based mainly on uniform loads, it is also wise not to store very heavy wares over a small area of a floor. Structural engineers design structures to with-

**Table 3.2 Minimum Uniform Live Loads (psf)**

| | |
|---|---|
| Apartment and hotel rooms | 40 |
| Assembly halls | |
| fixed seats | 60 |
| movable seats | 100 |
| Balconies | 100 |
| Bleachers | 100 |
| Corridors | 100 |
| (except 60 psf in apartment houses) | |
| Dance halls, gymnasiums | 100 |
| Driveways, sidewalks | 250 |
| Restaurants | 100 |
| Garages (passenger cars) | 100 |
| Libraries | |
| reading room | 60 |
| stacks | 150 |
| Manufacturing | 125 |
| Offices | 80 |
| Recreational areas | |
| (bowling alleys, poolrooms, etc.) | 75 |
| Skating rinks | 100 |
| Stairs | 100 |
| Stores | |
| first floor | 100 |
| upper floors | 75 |
| Theatres | |
| orchestra floor and balconies | 60 |
| stage | 150 |
| Warehouses | |
| light storage | 125 |
| heavy storage | 250 |

stand the expected loads, but owners must also be aware of the fact that unusual load concentrations may not have been part of the original design criteria. The occasional heavy loads due to some files or safes is considered in the design of office buildings by applying a concentrated load at a place where it most affects the structural element being designed. However, the possibility of having these heavy articles all over a floor is usually not considered. Some concentrated loads to be used in conjunction with uniform loads are listed in Table 3.3, taken from the BOCA Basic Building Code (BOCA [1970]).

The probability of having the full design LL on an entire floor clearly decreases as the area of the floor increases. To take into account this fact, most codes contain live load reduction provisions. One common method is to reduce the LL values for areas larger than 150 ft$^2$ at a rate of 0.08% per square foot up to a reduction of 60%. Since places of public assembly

**Table 3.3 Concentrated Loads (lbs)**

| | |
|---|---|
| Garages for passenger cars | 2000 |
| Garages for trucks | (not less than 150% maximum wheel load) |
| Office floors | 2000 |
| Sidewalks | 8000 |
| Stair treads (on center of tread) | 300 |

may receive the full loading, the reduction is not applicable to such structures. Similarly, it is quite unlikely that all or most floors of a multistory building would be fully loaded simultaneously; thus a reduction of the order of 20 to 50% is allowed in calculating the loads on columns that support several stories. The use of these reduction factors implies that in some very improbable cases, with the full design load acting on the entire tributary area, some structural components would be overloaded.

The dynamic effects of most live loads are not considered explicitly since a complete dynamic analysis would be excessively time-consuming. The occupancy loads are conservative enough to account for the increased stresses caused by the vibration of the structure, for example, as a result of people dancing. A suddenly released load (without a drop) produces maximum stresses that are twice those due to a slowly applied (static) force, but if the load is dropped from a height, the effect of impact is even greater. It is unlikely that the full design live load would act dynamically in a synchronized fashion.

Certain types of live loads may produce large dynamic stresses even though their magnitude may be small. In most instances the effects of possible vibrations are minimized by simple measures, in other cases complex dynamic analyses are necessary to calculate the dynamic response of the structure. For example, aircraft, suspension bridges, towers, elevator and crane supports are often analyzed for vibration effects; the methods of calculation are covered in books on structural dynamics.

The forces caused by rotating machinery and moving vehicles are dynamic. If a rotating flywheel of a machine is not precisely concentric with its shaft, or if it has an eccentric mass distribution, it exerts a periodic force on its supports. This force has the form $F\sin\gamma t$, where $\gamma$ is the frequency of rotation in radians/second. If the properties of the supporting structure (beam, slab) are such that its natural frequency is close to the driving frequency of the machine, large deflections and stresses may build up in the structure because of the resonance effect. This is similar to the effect when one jumps rhythmically on a flat board to excite it or when troops march in step on a bridge (which is, for this reason, forbidden).

Moving cranes in industrial facilities (Figure 3.2) and elevators are

**Fig. 3.2**
Industrial crane runway.

other examples involving dynamic live loads in buildings. To account for the dynamic effects of moving cranes, the static vertical lift-loads are increased by about 25% and, to account for the deceleration on the runway, a horizontal force equal to about 20% of the load is applied along the runway. The effective weight of elevators is usually increased by about 100% to include the dynamic effects produced by acceleration or braking.

Some cranes and supporting structures are designed for the stalling torque of the hoist motor, so that the hook (i.e., the load) cannot pull the structure down in cases of malfunction. The load equivalent of stalling torque is about three times the rated capacity. In an analogous situation, a sign plate may be fastened to its supporting tower in such a way that extraordinary wind gusts would blow the sign away without damaging the structure.

## 3.4 BRIDGE LIVE LOADS

The maximum stress in the members of a bridge depends not only on the weight of a moving vehicle but also on its position on the bridge. Thus, one has to determine the critical positions of the moving vehicles that produce maximum forces at various points along the bridge. This is usually done with the help of *influence lines* (see Volume 2, Chapter 16).

The question of maximum vehicle loading is very complicated for several reasons: (a) there are new truck-trailer combinations, (b) the effects of moving loads depend on the type of bridge and on its span, (c) the possibility of several heavy trucks passing critical positions on the bridge simultaneously must be considered, (d) moving loads cause vibrations and therefore higher stresses. Furthermore, maximum moments and maximum shears in the bridge occur under different load positions.

When traffic is held up, vehicles stand close together and the bridge receives the maximum static LL. It seems unduly conservative to populate a bridge only with heavy trucks, although this may be advisable near large plants where heavy trucks travel. On the other hand, the maximum dynamic effect occurs under fast traffic conditions, but then the spacing of vehicles is substantially greater and therefore the total weight is much less. The effect of the actual relative positions of heavy vehicles is significantly smaller on long bridges than on short bridges. A convenient simplification seems to be to assume a uniform lane load representing a "truck-train" on long bridges. On short bridges, however, the forces depend greatly on the size, weight, and position of each vehicle.

The probability of having simultaneous critical loading on several lanes is much smaller than critical loading of a single lane. Thus, a reduction of the total LL on a bridge more than two lanes wide is reasonable.

It is clear that bridge designers cannot investigate the stresses resulting from all possible loading conditions. Recognizing that the possibility of the simultaneous presence of several heavy trucks on short bridges is very small, highway design codes have developed standard loadings that simulate the effects of heavy vehicles. For a number of years during the last century, highway bridges were designed for a crowd of people at 80 to 100 psf. This loading was later made a function of the span, ranging from 60 psf (for very long spans) to 120 psf for short bridges (Waddell [1916]).

The loads recommended by the AASHO Specifications for Highway Bridges (AASHO [1969]) are shown in Figure 3.3. The two types of standard vehicles used for short span bridges are the simple H truck and the HS semitrailer. The first number denotes the total weight of the truck in tons, the second number (44) signifies the year of adoption of the loading (1944). One such vehicle is placed on the span. In addition to these individual truck loadings, bridges must also be investigated for a uniform lane loading combined with a single, concentrated load. This loading represents a train of trucks. It ranges from 320 lb/ft of traffic lane and 9000 lb force corresponding to the H 10-44 load, to a high of 640 lb/ft lane load and 18,000 lb concentrated force, associated with the HS 20-44 loading. The concentrated loads given are for determining moments; higher values are used for shear. These equivalent lane loads produce maximum stresses and deflections in long span bridges (greater

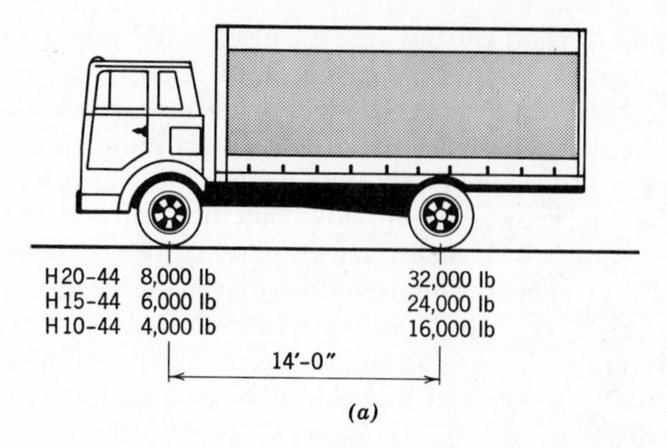


(a)

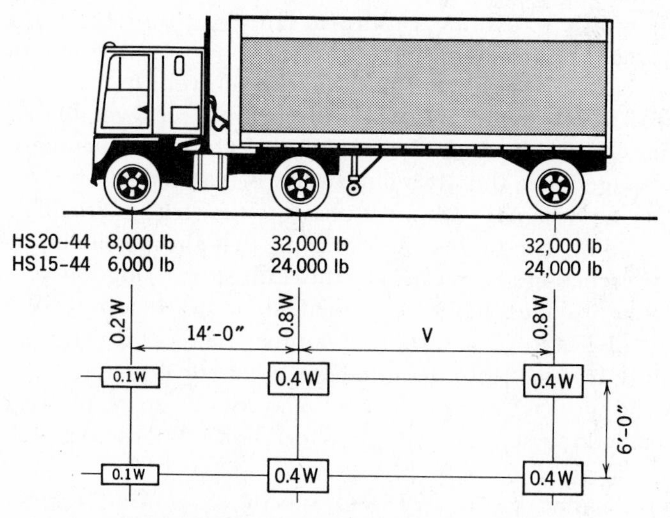


W = Combined weight on the first two axles which is the same as for the corresponding H truck.

V = Variable spacing—14 ft to 30 ft inclusive. Spacing to be used is that which produces maximum stresses.

(b)

**Fig. 3.3**
Standard truck loadings for highways. (*a*) Standard H trucks. (*b*) Standard HS trucks.

than 120 to 140 foot spans, depending on the type of stress or deflection). For highways that carry heavy truck traffic, the minimum recommended load is the HS 15-44. Bridge design specifications may also contain special design overloads for highway bridges that are to accommodate exceptionally heavy single vehicles under controlled conditions.

The specified AASHO loadings are designed to give about the same maximum shears and moments as the heaviest loadings permitted legally on the highways.

Maximum allowable loads are posted at some secondary highway bridges (you should observe and compare these limits as you travel). These limits are usually based on the evaluation of the safe capacity of the structure and, since the bridge may deteriorate, various local agencies rate the capacity of bridges periodically. The reserve capacity of most bridges and the conservatism in design is demonstrated by the fact that one sees trucks much heavier than the posted limits passing on bridges.

When a vehicle moves on a bridge, vibrations are set up because of the surface irregularities, the movement of the load, and the spring-mass interaction of the vehicle with the bridge. The result is that the live load stresses in the structure are increased. It has been customary, from the early days of bridge engineering, to account for this dynamic effect by an *impact factor*. This effect depends on a number of factors (such as speed, surface roughness of the pavement, characteristics of the vehicle, and span). The question of impact was debated at length during the 1890's because the lack of experimental data hampered the development of a rational estimate of the impact effect.

The first formulas accounted not only for impact but also for other effects that decrease with increasing span. The LL becomes a smaller portion of the total load (LL + DL) for longer bridges and therefore the fluctuation of stresses is also less. For the same reason, leading engineers recognized that the safety factor should decrease for long bridges. Consequently, impact formulas were inversely proportional to the length of the span $L$, decreasing rapidly for short $L$ and approaching a constant impact factor $I$ for long spans. The "average" of the formulas used by a number of leading bridge engineers in the 1890's was $I = 40{,}000/(L+500)$. This expression, correctly interpreted, was intended to account for all the dynamic effects of moving loads. A committee, established in 1907, conducted experiments on nearly 50 single-track railway bridges. They found that the unbalance of the locomotive drivers and the speed are important factors. In small spans the impact factor exceeded 60%. Waddell proposed the formula $165/(L+150)$ in 1912 to match the experimental data. For highway bridges he recommended the impact factor $100/(nL+200)$, where $n$ is the total width of the bridge divided by 20 ft. The inclusion of $n$ in the formula recognizes the decreased (less than proportional) impact effect for more than one lane of traffic and the effect of roadway width.

The presently used AASHO formula is

$$I = \frac{50}{L + 125}, \tag{3.1}$$

with a maximum of 0,3. It is obviously an oversimplification, but has been used successfully. Modern bridges tend to be shallower and therefore more flexible, which means that vibration effects and deflections are also greater. You can sometimes notice the vibration of bridges while driving slowly and letting other cars go by fast. Recent studies have indicated that the springing of vehicles and the vibration of the vehicle approaching the bridge are very important factors; under adverse conditions, mainly for spans less than 40 ft, the impact factor may reach or exceed unity. Extensive research on the dynamics of highway bridges is continually being carried on. It seems prudent to design highway bridges with some extra safety margins in anticipation of increasing traffic and heavier loads allowed by state legislation.

In addition to the vertical force resulting from impact, codes also prescribe two horizontal forces, one to account for the centrifugal effect on curved bridges, and another to approximate the braking force of a stopping truck (usually at 20% of the weight).

### 3.5 SNOW AND RAIN

Snow and rain loads affect the design of roofs. The design loads corresponding to the highest accumulation of snow in the various regions in the United States are shown on a map in Figure 3.4. These are minimum

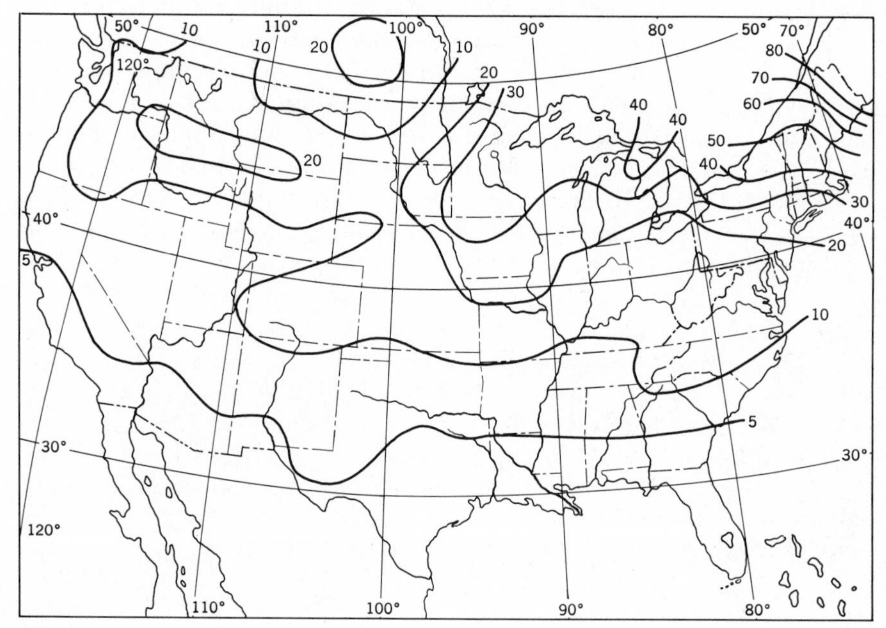


**Fig. 3.4**
Minimum snow loads, pounds per square foot.

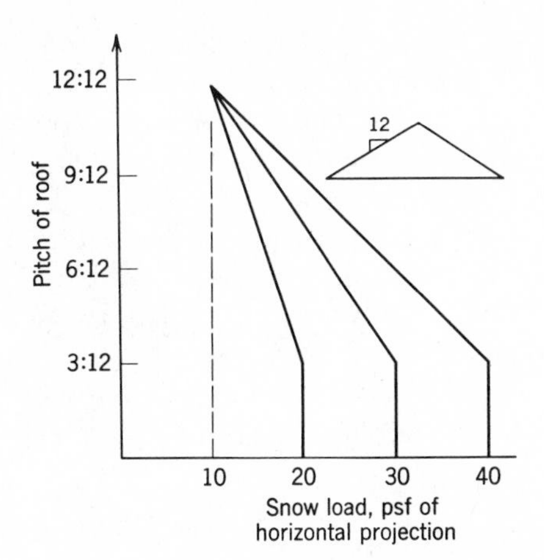


**Fig. 3.5**
Variation of snow load with slope of roof.

values and in certain areas, where exceptionally high, local accumulation of snow occurs (for example at high elevations), it is advisable to use larger values. One inch of snow is equivalent to about 0.5 psf load, but it can be more at low elevations where the snow is often denser. According to other conversion methods, compacted snow, which is 0.5 to 0.7 times the height of fresh snow, has a specific gravity of about 0.2.

The shape of the roof should also be considered. In some cases, drift or sliding causes buildup over sections of a roof. If the roof is steep, large accumulation of snow is not possible. Figure 3.5 shows an example for the reduction of snow load as a function of the slope of the roof. A minimum of 10 psf is recommended. For some structures snow may

**Fig. 3.6**
Heavy snow load.

accumulate in valleys of roofs or a partial snow load that covers only a part of the roof may develop because of rapid melting or wind action on the exposed surface. Such special conditions must be considered by the designer since partial loading may produce more critical stresses and deflections than loading over the entire roof. Several major roof structures have failed when subjected to unsymmetrical loading as a result of drifting snow. The probability of having maximum wind load and full snow load is generally small and negligible; this represents an additional consideration when we calculate forces due to snow and wind (see Section 3.10). Snow and ice loads of 60 psf or more on flat roofs are not uncommon in some areas. A design snow load of 400 psf was considered on Mt. Rainier at an elevation of about 5500 ft.

Rain load is not considered separately because it is usually less than the snow load and also because drains are provided on roofs where rain water might collect. However, failures have occurred when rain caused local deflections (either because of design or construction errors or clogged drains). When rain water collects, it may, under certain conditions, cause additional deflection which causes more water to accumulate. This progressive deflection and accumulation may continue and lead to failure (Chinn, Mansouri, and Adams [1969]). Such ponding failure occurs when the flexural stiffness of the roof system is small relative to the span. The AISC Specification (AISC [1970]) contains provisions on ponding that specify minimum stiffnesses for primary and secondary roof beams.

## 3.6 WIND AND BLAST LOADS

Wind and blast loads exert pressure (or suction) on the exposed surfaces of structures. Wind pressure is a very important design load in the case of tall structures. The magnitude, frequency, and distribution of wind loads depends on a number of factors and detailed information has been developed only in recent years. Code design provisions are, again, simplified to allow rapid design of most structures.

The wind forces are based on the maximum wind velocity 30 ft above the ground at the particular location. These velocities are assembled from weather station data (Figure 3.7). They are expected to occur only once in about 50 years. ASCE [1961] contains detailed information on the variation of the wind velocity.

The basic pressure loading resulting from the above wind velocity is

$$p = 0.002558 C_s V^2 \tag{3.2}$$

where $p$ is the pressure acting on vertical surfaces in pounds per square foot, $C_s$ is a shape coefficient, and $V$ is the basic wind velocity in miles per hour from Figure 3.7 (Thom [1968]).

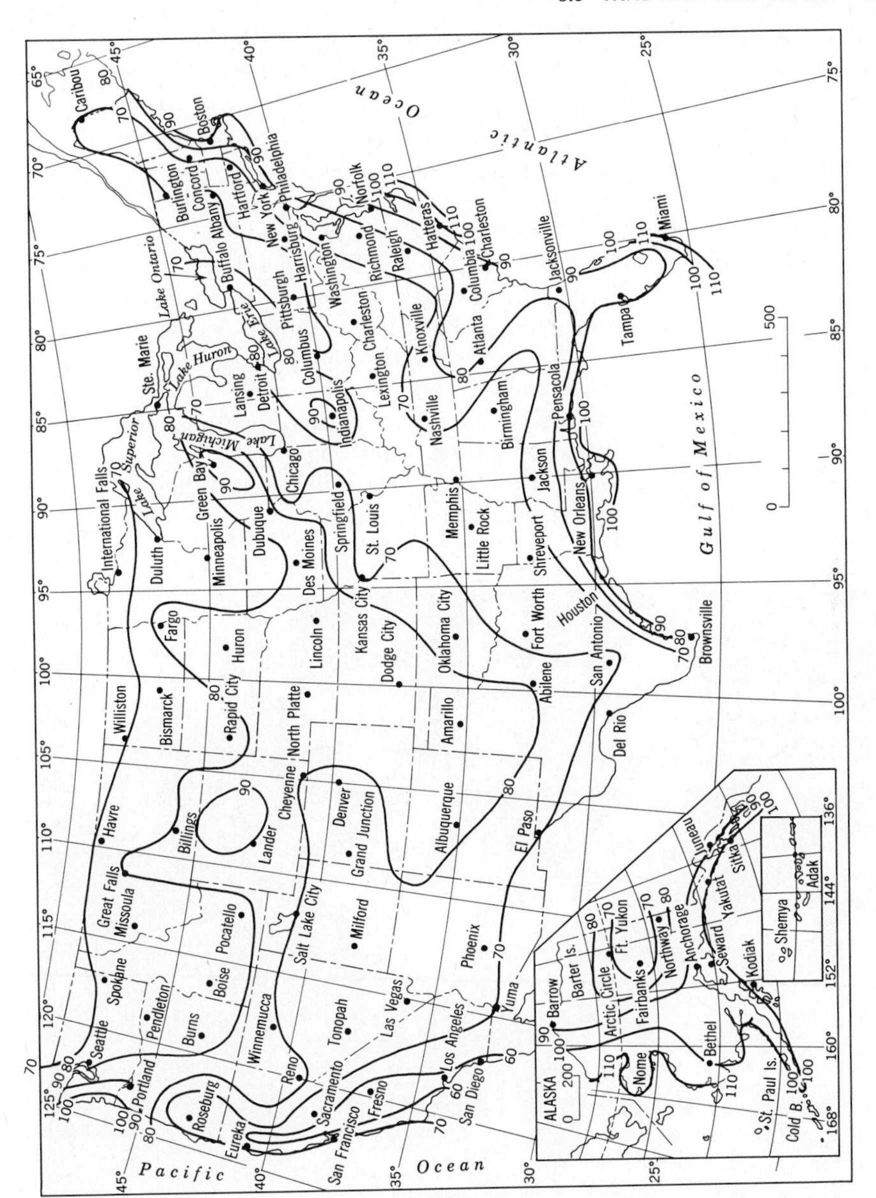


**Fig. 3.7**
Fastest mile wind velocities 30 ft. above ground at airports, 50-year mean recurrence intervals.

The coefficient $C_s$ depends on the shape of the structure, primarily the slope of the roof. For a box-type structure, the total shape coefficient is 1.3, of which 0.8 is for the pressure on the windward side and 0.5 is for the suction on the leeward wall. Detailed information on the shape coefficients obtained in wind-tunnel tests is contained in a task committee report (ASCE [1961]). Such coefficients provide us with a useful approximation to the actual distribution of pressure and suction.

The variation of wind velocity with height depends on the surface roughness of the terrain around the structure and on the temperature distribution of the interacting air masses. Most designers use the "seventh-root profile" law, according to which the velocity $V_z$ at elevation $h_z$ is $V_z = V_{30}(h_z/h_{30})^{1/7}$, where $V_{30}$ is the reference velocity at elevation $h_{30} = 30$ ft (Figure 3.7). If the terrain is rough (e.g., in urban areas), the increase may be much less and the exponent may be as high as 0.3.

Equation 3.2 does not take into account the rapid pressure variations and the effects of the roughness of the surrounding terrain. The effect of gusts

**Fig. 3.8** (a)
Hurricane winds. *National Oceanic and Atmospheric Administration*

may be considered in an approximate manner by using a gust factor $C_g$ in Equation 3.2. In the case of short gusts of about 1 sec duration, a factor of $C_g = 1.3$ is recommended; this is applicable to narrow structures. For longer gusts (10 sec duration), $C_g = 1.1$ may be used in the design of wide structures.

The BOCA Basic Building Code recommends 15 psf for short structures (less than 50 ft high), 20 psf for heights between 50 and 100 ft, and increasing pressure $20 + (h - 100)0.025$ for structures over 100 ft high, where $h$ is the height. Of this, 2/3 is pressure on the windward side and 1/3 is suction on the leeward side.

In areas where hurricanes, tornados, or cyclones occur, much higher wind forces should be used. Local codes usually contain guidelines, although not much is known about these forces.

Since the maximum wind forces on structures depend on peak pressure (gusts) rather than on average pressures, it is preferable to use a measure that reflects short variations in wind velocity. The quantity "fastest mile"

**Fig. 3.8** (*b*)
After Hurricane Camille. *PCA*

has been invented for this purpose; it represents the maximum velocity of a one-mile-long column of air passing the reference point. This is a good measure of the variation of wind velocity and design specifications in the future may use this quantity. A limited number of maps showing the fastest mile values are available.

Equation 3.2 has been used for a number of years. Recent extensive investigations of the dynamic behavior of structures showed that the determination of wind loads is a very complex matter and local surface conditions (surrounding terrain) are of primary importance. The flow of air is drastically changed when passing an object. The local velocity may increase by a factor of 1.5 in some practical cases. Wind tunnel tests on the models of the proposed structure together with all significant objects around the structure are necessary to study all factors. The wind tunnel investigation of the U.S. Steel Building in Pittsburgh is shown in Figure 3.9. In such a detailed study, in addition to the determination of forces, the maximum deflections and accelerations must also be evaluated to assure that occupants would not be subjected to discomfort for a long, cumulative time period. These investigations involve the study of local meteorological data and the statistical evaluation of structural response. Obviously, such elab-

**Fig. 3.9**
Wind tunnel test of the U.S. Steel building, Pittsburgh. *Boundary Layer Wind Tunnel, University of Western Ontario*

orate design procedures are justified only in the case of very special structures.

Wind forces may exert considerable uplift as exemplified by the action of airplane wings. You can demonstrate this by holding a sheet of paper at two corners and blowing above it. Uplift may amount to as much as 40 psf on common structures and must be considered in their design, especially if they are not supposed to fly. A number of light roofs and walls have failed as a result of wind suction or pressure.

Often brick or concrete block walls stand unsupported until the roof or floors tie them together. Surprisingly light winds may overturn such unbraced walls because of the large area exposed to the pressure. It is customary to use temporary cable or truss bracing during construction.

Relatively large amounts of unnecessary damage have been caused by wind pressure, quite apart from the destruction by hurricanes and tornados. Recent research studies have revealed the complexity of the subject of wind loading, yet much damage to ordinary structures could be avoided by paying attention to simple details.

The overturning effect of wind is an interesting question. If a slender structure such as a chimney is not anchored properly, it may topple over. Since strong gusts act only for very short time periods, it seems that overturning is usually not caused by gusts but by steady wind pressures of sufficient duration. Thus one uses a relatively small safety factor against overturning. It appears sufficient to assure that the overturning moment does not exceed about 70 percent of the stabilizing moment due to DL. Similar observations hold for the earthquake overturning effect.

Certain structures, such as transmission lines and suspension bridges, may receive periodic forces resulting from the von Kármán vortices. These occur when fluid flows past an object. The resulting vibrations have caused the failure of structures, the best known being the first Tacoma Narrows suspension bridge (see Chapter 2). Design methods have been developed to avoid resonance, limit "flutter," and assure adequate damping in such structures.

Blast loading involves a pressure phase, followed by a smaller suction wave (negative phase), as shown in Figure 3.10*a*. Design information is available to aid in the design of structures subjected to blast loading for detonations of various yield and for various distances from the blast (Newmark [1956]). The pressure resulting from sonic boom is only about 2 to 3 psf, however its effect is magnified by a dynamic factor of about two because of the very rapid variation of the pressure (Figure 3.10*b*). The damage caused by sonic boom is restricted to glass breakage and minor plaster cracking, except for low flying supersonic aircraft, in which case the pressure can be several times greater (Sharpe and Kost [1971]).

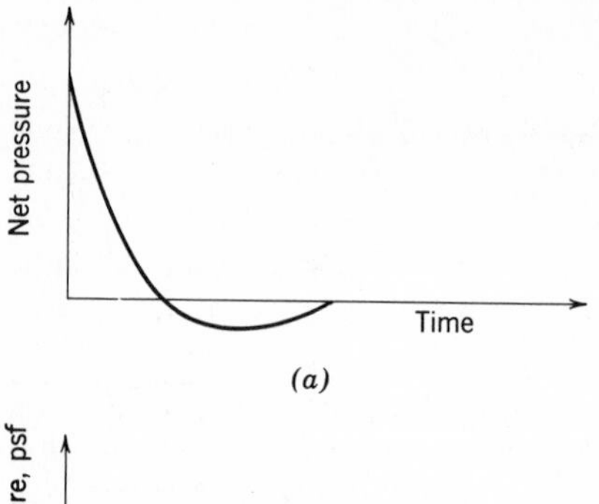


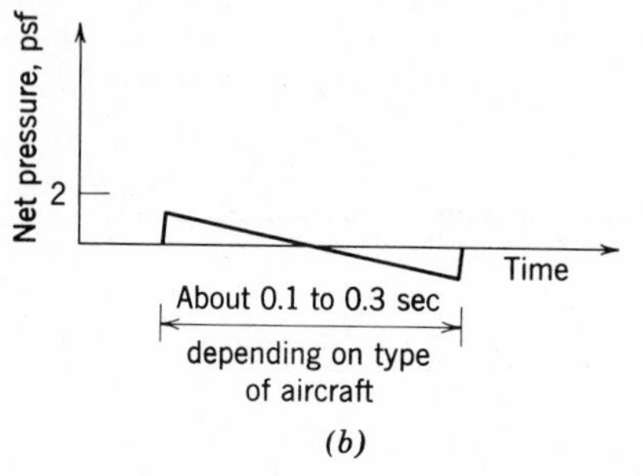


**Fig. 3.10**
Blast and sonic boom loading. (*a*) Blast loading. (*b*) Pressure resulting from sonic boom.

## 3.7 EARTHQUAKE LOADS

Earthquake loads affect the design of structures in areas of great seismic activity, such as in California, Alaska, and Japan. The highly irregular (random) shaking of the ground transmits accelerations to structures and the mass of the structure resists the motion due to the inertia effects. The total inertia force (usually equal to the horizontal shear at the base of the structure) ranges from about $0.03W$ to $0.1W$ or more for most buildings, where $W$ is the total weight.

The response of structures to earthquakes depends on a number of factors: the characteristics of the ground motion, the stiffness and mass of the structure, the subsoil conditions, and the amount of damping. A complex dynamic analysis for a specific (measured or artificially generated) ground motion is possible with the help of computers. This is necessary in the case of special or unusual structures. However, for the design of the majority of common types of structures, simple methods have been developed.

The seismic provisions of most codes are taken from the latest revision of the "Recommended Lateral Forces" published by the Structural Engineers' Association of California (SEAOC [1968]). However, numerous local (state or city) codes do not have proper earthquake design requirements. The maximum dynamic forces are approximated by equivalent static forces. The total horizontal force, or base shear $V$ is

$$V = \mathrm{ZKCW} \tag{3.3}$$

where $Z$ is the zone factor that depends on the seismic intensity, $C$ is a function of the flexibility and the mass of the structure, and $K$ represents the ductility in various types of structures. There are four seismic zones in the United States as shown in Figure 3.11; these are based on the history of seismic activity in the various regions (Algermissen [1969]). It is seen that the value of $Z$ varies widely from zone to zone. Such a map is, obviously, rather crude but more detailed subdivisions will be possible only when more records and seismological surveys become available.

The lateral force coefficient $C$ depends on the fundamental period of vibration ($T$) of the structure. $T$ is the time interval necessary for the structure to complete one vibration when released from a deflected position corresponding to the fundamental mode shape. It depends on the mass and stiffness properties of the structure and on the soil conditions. $T$ ranges from about 0.1 second for a low, stiff building to about 5 seconds for a tall, flexible skyscraper. For most structures, $T$ is in the 0.2 to 0.4 second range. It may be determined from detailed vibration analyses of the structure, or in most cases

$$T = \frac{0.05H}{\sqrt{D}} \tag{3.4}$$

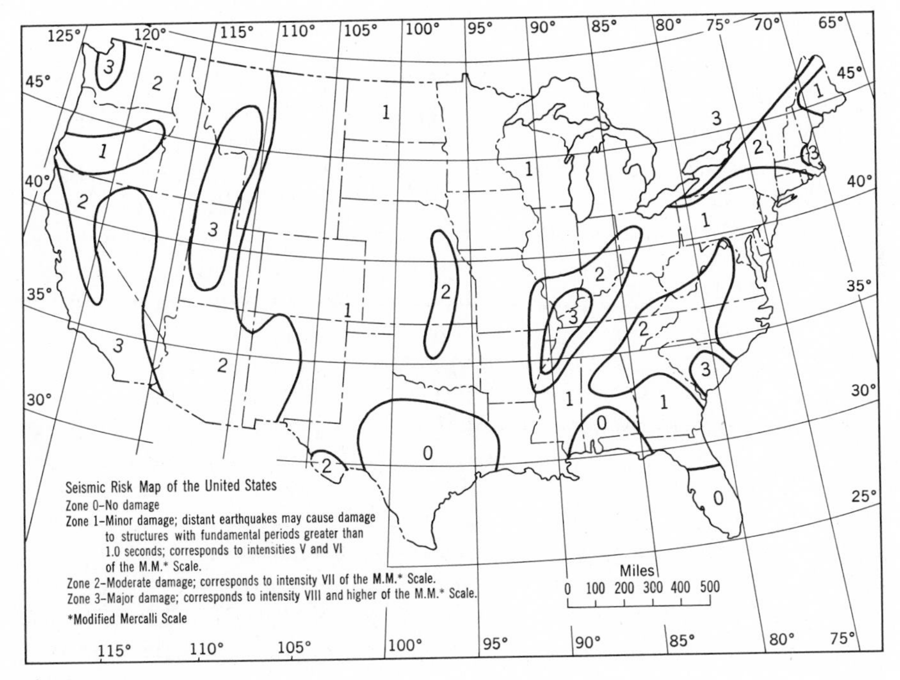


**Fig. 3.11**
Seismic risk map.

where H and D are the height and depth of the building in feet, respectively.

The coefficient $C$ is then given by the formula

$$C = \frac{0.05}{\sqrt[3]{T}} \leq 0.10 \qquad (3.5)$$

The forces on a structure also depend on the type of construction (K factor). As examples, $K$ is specified as 0.67 for an unbraced ductile frame, and 1.33 for a boxlike structure. Rigid elements, such as masonry walls, tend to attract higher forces. Extreme care must be taken to avoid brittle failures. It is highly important to assure that all structural (and nonstructural) elements have adequate deformation capacity (ductility). The SEAOC recommendations contain further guidelines for earthquake-resistant design. In the case of unusual structures such as very tall buildings and nuclear reactor facilities, complex analyses must be performed.

## 3.8 WATER AND EARTH PRESSURE, ICE AND WAVE LOADS

The pressure exerted by a fluid is normal to the surface of a submerged object. The magnitude of the pressure is

$$p = \gamma h$$

where $\gamma$ is the unit weight of the liquid and $h$ is the elevation difference between the liquid surface and the point in question. This linear pressure distribution occurs in tanks, vessels, and underwater structures.

Structures below ground such as foundation walls, retaining walls, or tunnels are subjected to earth pressure. This pressure depends on many variables, such as the cohesion and friction in the soil, the possibility of swelling (which may occur in clays), and the rigidity of the structure.

For cases in which the structure can deflect slightly under the soil pressure, the soil will "follow" it, thus developing internal friction that reduces the effective force on the structure (reaching a state of "active pressure"). In such cases the soil pressure is considerably reduced.

Vertical pressure in the ground is the unit weight (90 to 120 lb/ft$^3$) multiplied by the depth to the point in question. Lateral soil pressure is reduced from this amount by cohesion and friction; the amount of reduction ranges from about 40% for sands to as high as 80% for compacted soils or cohesive clays. In many simple design situations, a "liquid" pressure of about 30 psf per foot of depth may be used for dry soils. The pressure due to water is an additional 62.4 psf per foot of depth below the water level. Both of these loadings have triangular pressure distributions.

It is customary to install drain pipes next to retaining and foundation walls to reduce the water pressure. Buoyancy forces resulting from the presence of water are significant, for instance, in the case of large hollow tubes (penstocks) that are buried between reservoirs and power stations. Below-grade swimming pools sometimes "pop out" if the pool is emptied and the surrounding water table is high.

Ice may form on surfaces and this should be considered when designing slender members such as cables or towers. The formation of ice increases the weight and also the surface area on which the wind pressure acts. An ice layer of 1- to 2-in. thickness is commonly used, although in some locations this value may be much higher. A very different type of ice loading is encountered on bridge piers where river ice may occur. The force on a bridge pier caused by a mass of ice floating down a river can be formidable; its magnitude is often estimated to be equal to the area of ice in contact with the pier multiplied by the compressive strength of

**Fig. 3.12**
Ice on microwave tower, British Columbia.

*B. C. Jennings*

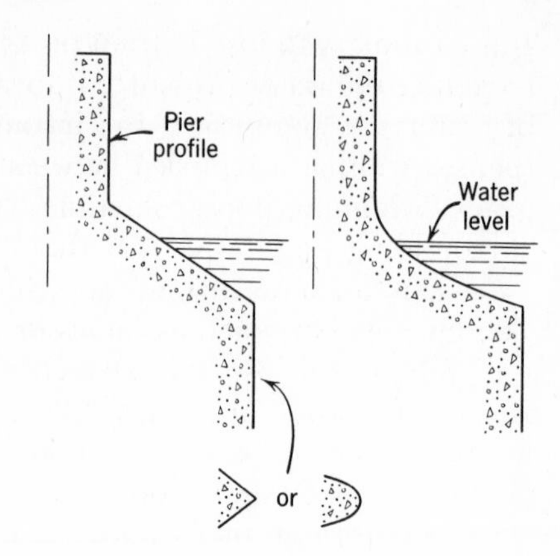

**Fig. 3.13**
Icebreaker on bridge piers. *Hardesty & Hanover and N.Y. S. Thruway Authority*

the ice. This force may be reduced by shaping the upstream face of the pier to reduce the exposed area or to force the flowing ice to rise up on a ramplike edge that breaks the ice (Figure 3.13).

Wave loading affects the design of off-shore structures and harbor facilities. The forces exerted by hurricane waves on surface structures and by fluid motion on submerged structures is a highly complex subject (Quinn [1961]). Substantial design information has become available only in recent years.

The distribution of forces on certain submerged bodies (e.g., cylindrical shapes, such as round piles) subjected to idealized flow conditions has been analyzed. The purely analytical methods are usable only in some cases; usually some experimental data is used in the prediction of wave loads. The direct wave effects on harbor facilities must be approached on a probability-statistical basis to relate hurricane frequency, maximum observed wave heights and speeds, and bed surface properties. Design curves are being developed for such analyses, but we expect significant improvements in this area.

Extensive studies of hurricane and wave forces in the Gulf of Mexico have yielded useful information for the design of off-shore oil drilling platforms. Maximum hurricane velocities are about 120 mph and maximum wave heights are roughly 40 ft, although 60-ft high waves formed during a storm in 1945. Extensive statistical studies and good engineering judgment are required in the determination of design wave velocities and heights.

## 3.9 LOADING INDUCED BY TEMPERATURE, SHRINKAGE, AND LACK OF FIT

Forces may be created in structures when relative movements between various points in the structure are not free to occur. Such relative movements may develop as a result of temperature changes, shrinkage, or settlement of some of the supports. These loads should be considered in design because surprisingly large forces may result in structural components where unrestrained expansion or contraction cannot occur.

**Table 3.4 Coefficients of Thermal Expansion, $\alpha$**

| $\times 10^{-6}$ per F° | |
|---|---|
| Aluminum | 12.8 |
| Brickmasonry | 3.4 |
| Concrete | 5.5 |
| Copper | 9.3 |
| Mild steel | 6.5 |
| Plastic | 40. |
| Stainless steel | 9.9 |
| Timber | 2.0-3.0 |

As a simple example, take a steel bar and assume that the two ends are held fixed. If a temperature drop of only 50°F occurs over the entire length of the bar (e.g., if cold liquid flows past the bar), the strain in the bar is

$$\epsilon = \alpha \Delta T = 6.5 \times 10^{-6} \times 50 = 325 \times 10^{-6}$$

where $\alpha$ is the coefficient of thermal expansion (see Table 3.4). The stress is

$$f = E\,\epsilon = 29.5 \times 10^{6} \times 325 \times 10^{-6} = 9600 \text{ psi}$$

where $E$ is the modulus of elasticity of steel. Note that the length and the cross-sectional area of the bar do not matter; hence, it is not possible to decrease temperature stresses by increasing member sizes. Temperature stresses also develop if materials with appreciably different thermal coefficients (e.g., mild steel and stainless steel) are joined together. The nearly equal thermal expansion of steel and concrete makes it possible to use steel as reinforcement in concrete members. Thermal loadings that produce large forces and stresses occur in nuclear reactor structures and in other types of industrial facilities. Shrinkage, which occurs in concrete, has a similar effect to temperature change.

In the design of large structures it is customary to design expansion joints to allow the structure to "breathe" under temperature variation or shrinkage without straining and cracking the structure.

Stresses may also result from lack of fit, that is, when a member of improper size is forced into place during construction. This problem is usually avoided by careful detailing of connections. It should be noted that lack of fit, temperature, and shrinkage stresses do not develop in statically determinate structures.

## 3.10 LOAD COMBINATIONS

Engineering judgment must be exercised when determining the critical load combinations. It is not necessary to superpose all maximum loads. For example, since the probability of their joint occurrence is negligible, the maximum wind loads are not combined with the maximum snow or earthquake loads. The critical combinations are usually specified by codes. The logical method would be to base load analysis on a consistent statistical and probabilistic approach, but the information needed to be "logical" is far from being available today.

In recognition of the highly unlikely occurrence of maximum wind or earthquake loads simultaneously with the full value of other live loads, codes generally allow a 33% increase in allowable stresses under these load combinations. A notable exception is in nuclear reactor design, in which case the live loads are not as random and variable as in other structures and also the consequences of failure may be extremely grave; therefore, the 33% increase is not allowed.

Larger safety factors against failure are required for poorly defined live loads than for dead loads. For this reason it is also customary to group the various loads according to the safety factors associated with them. The probability of critical load combinations and their effects on the safety of occupants and the economic loss of failure varies greatly from structure to structure.

## 3.11 SUMMARY

The determination of loads is an important step in the design process. In the case of unusual or special structures, the load analysis may require extensive study. Various codes contain recommended loads for routine cases that have been used with success. In any case, the structural engineer must ascertain the expected loads, their relative importance, and the associated structural performance.

### *Suggested Reading*

McGuire, W. [1968]: *Steel Structures,* Chapter 3, Prentice-Hall, Englewood Cliffs, N.J.

## PROBLEMS

3.1 Sketch a typical frame of a short multistory building (e.g., your classroom building or apartment house); give approximate dimensions and show all loads acting on it, including values.

3.2 What is your estimate of the actual maximum LL: (*a*) in your class room, (*b*) in the corridor?

3.3 A circular concrete chimney is 50-ft high and its inside diameter is 80% of its outside diameter ($d_i = 0.8d_o$). How large must $d_o$ be to avoid tension in the concrete due to wind load? What is the overturning moment for $d_o = 8$ ft? Sketch a foundation that can counteract this moment. $C_s = 0.5$.

3.4 Compare the LL on a bridge resulting from full loading by "average" highway traffic and by a crowd passing through.

3.5 Demonstrate the effects of impact by releasing or dropping a weight on a beam made of a slender rod.

3.6 A 1-in. diameter aluminum rod is placed inside a steel tube with an inside diameter of 1.25 in. and a wall thickness of 0.25 in. The rod and the tube are connected at their ends. What are the stresses caused by a temperature drop of 50°F? How can we reduce these stresses?

3.7 Give examples where thermal effects are important.

3.8 Determine the legal maximum truck loads permitted in your state, and compare with the AASHO HS 20-44 truck.

3.9 The impact formula for bridges includes *L,* the length of span, as the only variable. What other variables must affect impact?

3.10 A parking garage is being designed for ordinary automobiles. Using a basic floor size of 60 ft by 100 ft, and allowing sufficient room for movement of the autos in and out of the parking area, determine the total expected weight of autos on one floor. What is the corresponding average uniformly distributed load? Compare with code values.

3.11 A swimming pool, 20 ft wide by 40 ft long and 10 ft deep, is built of reinforced concrete, with 10-in. thick walls and a 12-in. thick bottom. After the pool is constructed, back-filled externally with sand, and filled with water, the ground water level rises to within 4 ft of the surface. Is it safe to empty the pool? If not, how far down can it be emptied without danger of the pool popping out of the ground?

3.12 It is proposed to use one room of an office building for storage of boxes of old correspondence, orders, and other paper. The individual boxes will be stacked to a depth of about 5 ft. If the room has been designed for a 100 psf live load, is the proposal satisfactory?

3.13 Determine the magnitudes of snow and wind loading specified in the building code for your community. Do the values seem to be in line with the values indicated in the text?

**CHAPTER 4**

# Structural Form

City Hall, Toronto, Canada.
*Architect: Viljo Revell, Helsinki; Associate architects and engineers: Searle Wilbee Rowland, Toronto*

# 4

# Structural Form

Man-made structures are utilized to house and support the activities of man. The first function, that of housing man's activities, requires the creation of well-defined, enclosed spaces. These spaces range in size and complexity from modular housing units to 100-story office buildings, from schools to large arenas and convention centers, and from jumbo jets to manufacturing plants.

Space is usually defined and enclosed by an architectural system imposed on a structural system. As an example, consider a high-rise office building consisting of light-weight exterior curtain walls and movable interior partitions fastened to a multistory structural frame. The frame must resist all applied loads including the weight of the walls and partitions, the live load, and the effects of the external environment. The floor system, which is usually a combination of structural and architectural components, completes the subdivision of interior space.

The second basic function of man-made structures is in facilities built to support and encourage man's innumerable activities. These support facilities include containment of material (water and petroleum tanks), providing a passageway for vehicular movement (highway and railway bridges), modifying the natural environment (dams), supporting a single load at a fixed point (television transmitting antenna), and creating usable energy (nuclear power generating stations). Most of these structures offer little or no opportunity for architectural embellishment and accordingly must have their architectural aspects integrated directly into the structural system.

We define the part of the structure that must resist the applied loads as the *structural system*. The ultimate role for any structural system is that of transmitting forces through space from the source of load to the foundation. The technical success of the system depends on how efficiently this flow of load is accomplished.

The mode of force transmission is a function of the geometrical configuration *(form)* of the load resisting structure. We shall examine the basic types of structural form in this chapter, presenting the fundamental factors pertinent to structural form and discussing typical forms for a variety of structural situations. We cannot overstress the importance of this topic, for the decision making leading up to the final geometrical form for a given structure represents the highest level of structural engineering practice. Structural form is also treated subsequently in a more specific and quantitative manner in Volume 4, where we shall rely on your having achieved a clear understanding of the many aspects of structural system behavior treated in the first three volumes.

As a vehicle for discussing structural form, consider the elementary problem of supporting a weight $Q$ at the center of an open span of length $L$, between points $A$ and $B$ (Figure 4.1*a*). The problem seems simple—to devise a structure that will transmit the point load $Q$ to foundation points $A$ and $B$. We shall discuss several possible solutions, but you should realize that there are an infinity of choices.

A second problem to be considered is that of a uniformly distributed loading $q$ over a rectangular area (Figure 4.1*b*). The structural requirements for a surface loading are more demanding than for a single load as the structural system must provide the basis for closing in or covering the area. We should realize that the surface loading is by far more prevalent than the single concentrated load case.

**Fig. 4.1**
Loading situations.

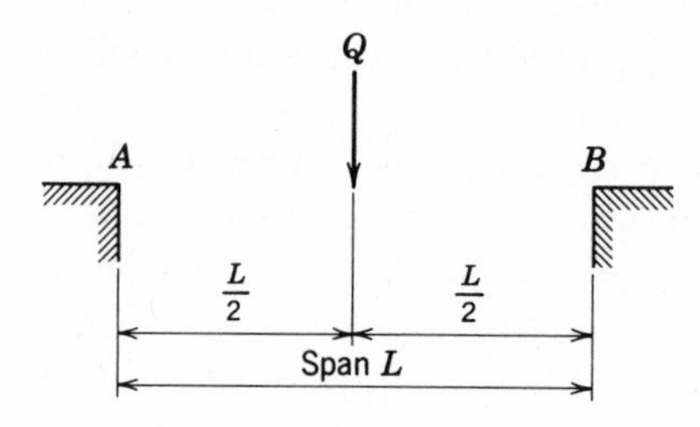


(*a*) Concentrated load.

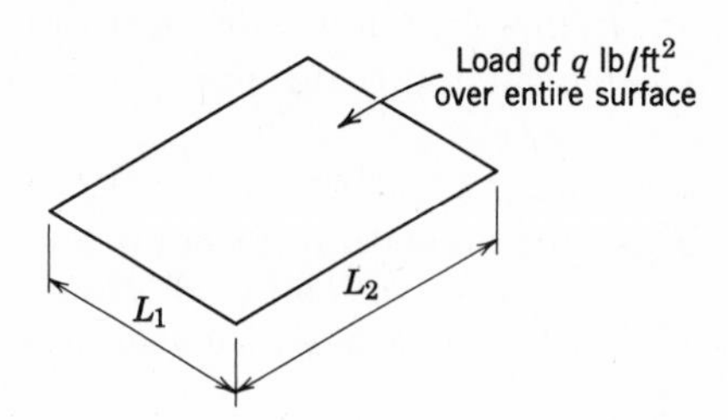


(*b*) Distributed etc.

## 4.1 TENSION OR COMPRESSION STRUCTURES

### Tension Structures

Some insight into man-made structural forms is gained by observing various forms occurring in nature. The web of a spider (Figure 4.2), exceedingly strong for its weight, is probably one of the most efficient tension structures ever built. An element of the web has essentially no bending stiffness and no resistance to compression; therefore the elements must be so arranged that the imposed loads (spider and prey) can be carried by tension alone. The configuration of the web changes continuously as the spider changes position, creating the shape required to support the changing load.

**Fig. 4.2**
Natural tensile structure.

*E. M. Rattensperger. Dept. of Entomology, Cornell University.*

Returning to the concentrated load case, it is rather obvious that we can support the load as shown in Figure 4.3 by a flexible cable attached at points $A$ and $B$, provided the foundations can resist the forces. This tensile structure transmits the load to the foundations by pure tension in the cable. The summation of vertical components of cable tension must bal-

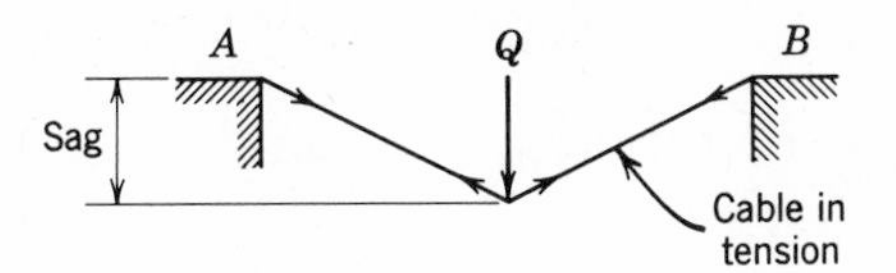


**Fig. 4.3**
Tensile structure.

ance the applied load, and the cable must have a finite value of sag in order to carry the load. The cable shape providing equilibrium with no bending is called a *funicular curve* (derived from the Latin *funiculus,* "small rope"). The flow of forces to the support points shows that this type of structure will always exert both vertical and horizontal forces on its supports.

Suppose the position of the load is changed—what happens? The cable geometry changes, taking on a new funicular curve. The unique form of the cable for each load position illustrates the basic dependence of the structure on its form. Engel [1967], in his book on structural systems, aptly classifies the tension structure as form-active. Provided we have a material with high tensile strength available, such as steel, the form-active tensile structure is an efficient solution to the problem.

Tensile forms for several different planar loading conditions are summarized in Figure 4.4. The funicular shape for a series of concentrated loads will always be a series of connected line segments. The shape is readily determined by a simple experiment with a string and a series of weights (see Problem 4.1). As the number of equal loads becomes infinite, or as we go to a uniformly distributed loading, the resulting shape is a parabola (Figure 4.4*c*). The true elegance of a simple tensile structure is best shown in a suspension bridge such as the Walt Whitman Bridge in Figure 4.5.

An obvious method for extending the planar cable to cover an area and support a distributed surface loading is to use a series of parallel cables, side by side, with the roofing system spanning between them (Figures 4.6*a* and 4.7*a*). Another variation is a doubly curved surface (Figure 4.6*b*) formed by using two systems of cables and curving one set upward and

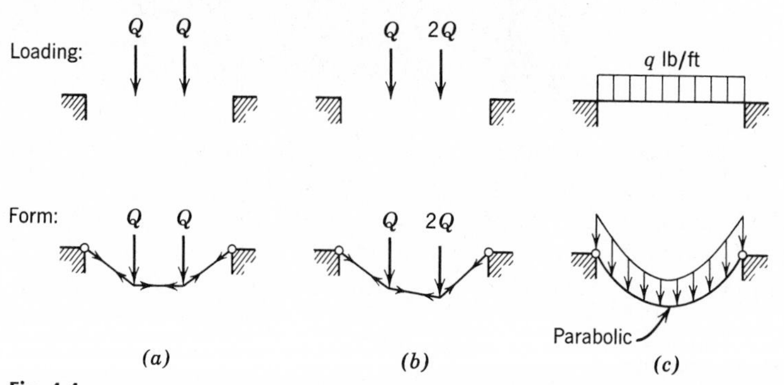


**Fig. 4.4**
Planar tensile structures. (*a*) Equal concentrated loads. (*b*) Unequal concentrated loads. (*c*) Uniform distributed load.

**Fig. 4.5**
Walt Whitman Suspension Bridge. *Bethlehem Steel*

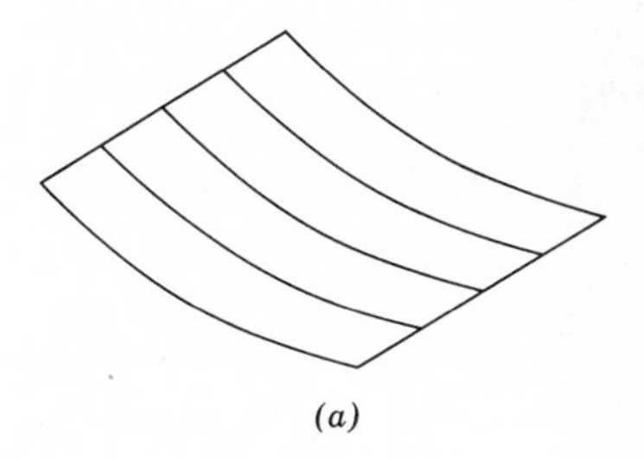



**Fig. 4.6**
Surface tensile structures. (*a*) Parallel cables.
(*b*) Orthogonal cables (saddle shape). (*c*) Radial cables.

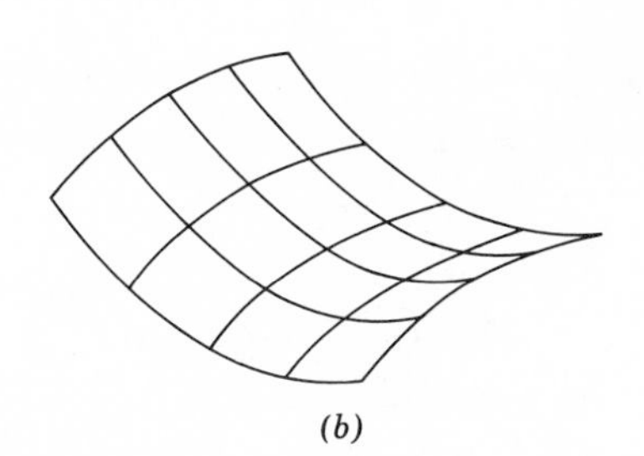



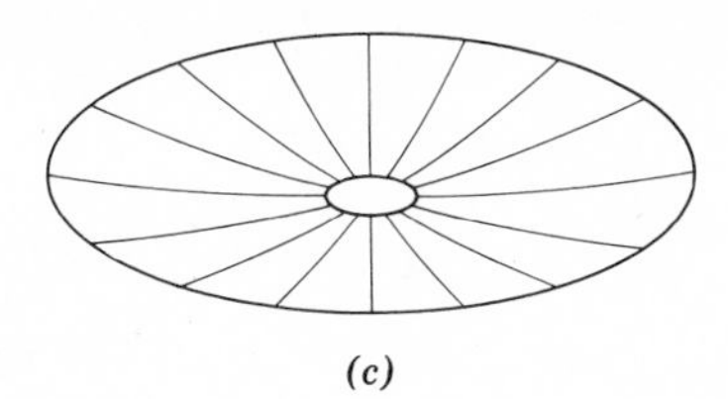

**Fig. 4.7**
(*a*) Dulles Airport Terminal, Washington, D.C. (*b*) Madison Square Garden during construction, New York.

*Bethlehem Steel*

the other downward. A radial system of cables is often used when the plan area to be covered is circular (Figures 4.6*c* and 4.7*b*). Striking architectural structures are made possible by imaginative use of cable networks.

Summarizing, a *tension structure* redirects the applied loading to the supports by tension alone in the primary members of the structure. It requires a construction material with high tensile load capacity. The shape of a tensile structure is a unique function of the magnitude and position of the applied loads.

## Compression Structures

There are two simple structural forms suited for carrying forces by compression alone—the *column* and the *arch.* A third form, the *shell,* will be discussed subsequently. A column is a straight member loaded along its centroidal axis with a compressive load. Except when it is extremely short, the column is less efficient than a tensile member because it has the tendency to buckle when compressed. The inclined struts in the Solleks River Bridge (Chapter 1) are columns.

The arch form is introduced by considering again the loading case of the single concentrated load (Figure 4.1*a*) and the tensile structure solution (Figure 4.3). If the cable geometry is inverted as shown in Figure 4.8, preserving the funicular curve, the two members supporting the load act in pure compression. This type of compressive structure falls within the general category of arches.

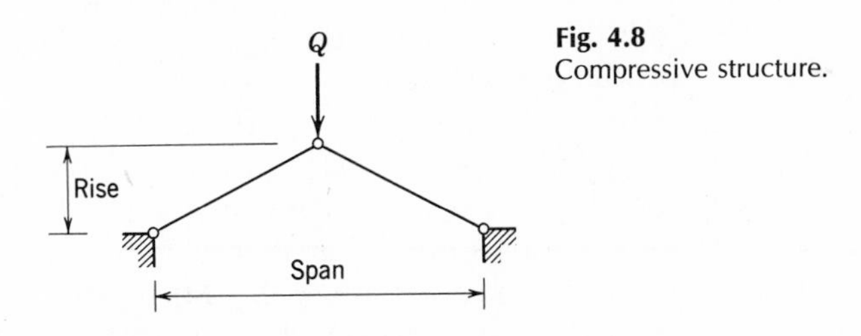


**Fig. 4.8**
Compressive structure.

The numerous natural rock arches existing in certain areas have been produced by the erosive forces of nature acting on large rock masses (Figure 4.9). They continue to stand only because the dead load stresses are primarily compressive in a properly shaped arch. The principle of compressive arch action has been utilized in masonry construction for centuries; in fact, the arch is the only feasible form for large structures made completely from a material that is weak in tension, such as unreinforced masonry.

The arch does not have the geometric flexibility inherent in a cable

**Fig. 4.9**
Rainbow Bridge National Monument (278-ft. span, 309 ft. high).

*Utah Tourist and Publicity Council*

because the cross section of a compressive member must have much more massive proportions than a cable in order to prevent buckling under compressive stress. As the loading deviates from that which dictated the original shape of the arch, the arch retains its original shape but cannot transmit the loading to the supports by compressive stress alone. It will thus undergo bending action in addition to axial compression. A well-designed arch will have as little bending as possible under the combination of loads for which it is designed.

Compressive (arch) forms for the several planar loading conditions in Figure 4.4 are obtained by inverting the tensile forms shown there, with the resulting structures of Figure 4.10. The flow of load through the members of the arches to the foundations is easily visualized. The multiple-span, reinforced concrete arch bridge over the Fox River in Illinois

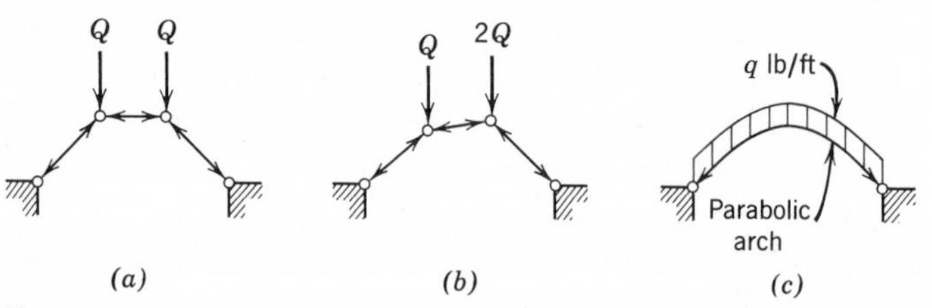


**Fig. 4.10**
Planar compressive structures. (*a*) Equal concentrated loads. (*b*) Unequal concentrated loads. (*c*) Uniform distributed load.

**Fig. 4.11**
Multispan Arch Bridge, Fox River, Ill.

*PCA*

utilizes arch spans the full width of the roadway (Figure 4.11). We should compare the simple arches of Figure 4.10 with those of the great Swiss engineer, Robert Maillart, who designed many elegant arch bridges in the first four decades of the 20th century. The grace and poetry of Maillart's many structures are described in rich pictorial form by Bill [1969].

The Eads Bridge (Figure 4.12) across the Mississippi River at St. Louis

**Fig. 4.12**
Eads Bridge, St. Louis, Mo.

**Fig. 4.13**
Compressive structures.
(*a*) Parallel arch structure, Island Garden Arena, West Hempstead, N.Y.

*AITC*

(*b*) Radial arch structure.

*Wood Preserving*

is another outstanding arch structure. The bridge has three arched rib spans of about 500 feet each. It was completed in 1874 and has been in continuous use since then, carrying both vehicles and trains on its two decks. The Eads Bridge was America's first major steel bridge and had span lengths that were completely unprecedented at its time of construction. This beautiful structure is now a National Historic Monument. The nearly insurmountable political, commercial, and technical obstacles faced by Eads should be studied by all budding engineers (Steinman [1957] and Gies [1963]).

Adapting the arch form to carry a surface loading can be done by either using a series of arches side by side to form a structure with the shape of a cylinder (Figure 4.13*a*) or by arranging them in a radial form to take on the shape of a dome (Figure 4.13*b*). In either case, the forces flow through the arch members, by compressive stress, to the base of the structure. A further refinement, that of eliminating the individual arches and replacing the entire system with a continuous curved surface, will be discussed subsequently.

All arch forms exert large outward horizontal forces at their bases. In single-span arches, these forces must be resisted by the supporting foundations or by a tension tie extending from one base to the other; the latter case is discussed in the following section. In multispan arches (Figure 4.11) the horizontal thrusts at the interior supports tend to cancel each other when all spans have similar loadings, but the problem still exists at the two outermost foundations.

In summary, the ideal *arch* form for a given loading is the inverse of the funicular shape. It can be built from materials that have high compressive strength, and it must have sufficient stiffness to prevent buckling.

## 4.2 TRUSS FORMS

Consider again the problem of supporting a single load $Q$ at the center of a span of length $L$.

The horizontal forces exerted on the foundations by a cable or an arch can be eliminated by adding a member that spans directly from $A$ to $B$, and then placing one of the ends of the new structure on a roller (Figure 4.14). If all joints are hinged, the additional member for a cable structure

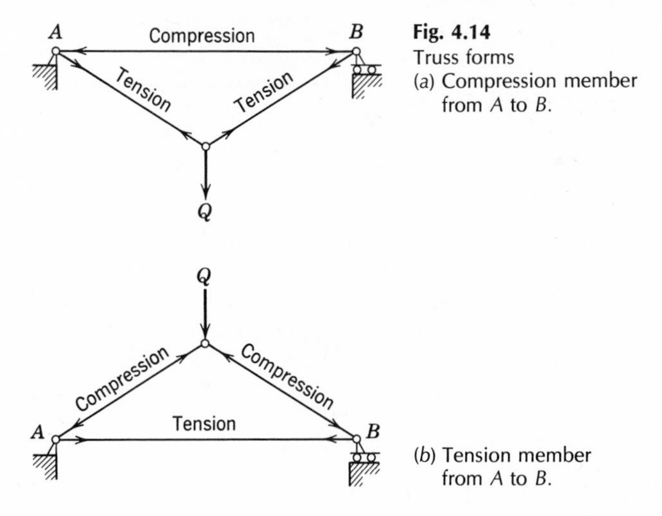


**Fig. 4.14**
Truss forms
(*a*) Compression member from *A* to *B*.
(*b*) Tension member from *A* to *B*.

will be in pure compression, while for the arch it will be in tension. We thus have a triangular form with both tensile and compressive elements, called a *truss*. The force transmission is provided by redirection of the loads into a series of forces in equilibrium with each other and any external loads at each hinged joint. The truss always consists of a number of straight components with roughly half in tension and the other half in compression. Any number of geometrically rigid triangles can be interconnected to give a stable configuration. Suitable truss forms for carrying various types of planar loads are given in Figure 4.15.

We should note that in contrast to the cable, the truss does not have to change shape under different loadings. Furthermore, if loads are applied only at the joints, it does not have the undesirable bending found in an arch subjected to variable loads. The only quantities that vary with loading are the magnitudes of the member forces. The assemblage of members is capable of redirecting any conceivable system of joint loads to any proper set of reaction points. Finally, the number of truss members and their precise configuration are design variables that give great flexibility to the engineer in shaping the structure to fit the specified problem at hand.

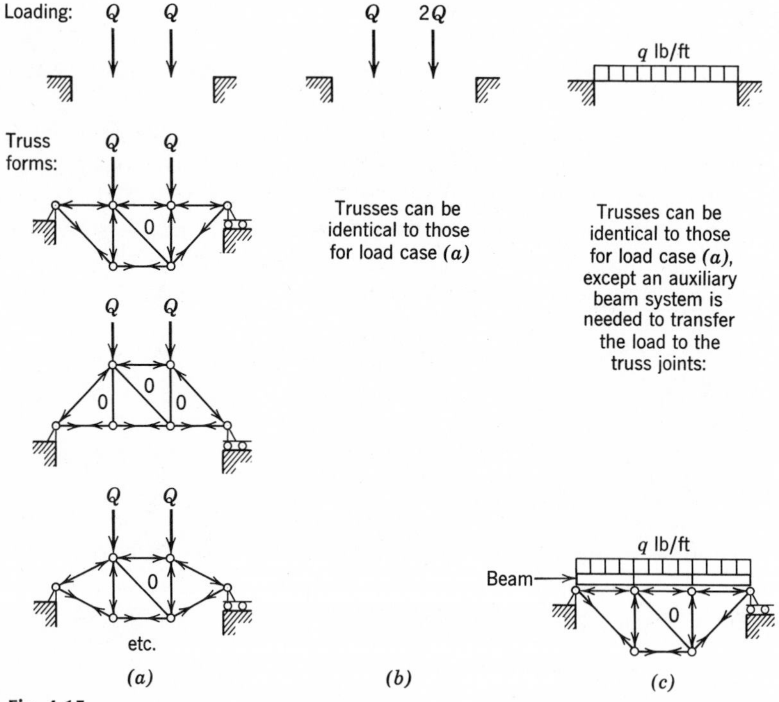


**Fig. 4.15**
Truss structures.

**Fig. 4.16**
IBM building, Pittsburgh, Pa.

*AISI*

The IBM Building in Pittsburgh (Figure 4.16) has four trussed walls on its exterior which tend to act as the faces of a large cantilevered tube when the building is subjected to wind loads. Loads resisted by the wall trusses include the wind effects as well as about half the total floor loading of the building. The steel trusses are sheathed in stainless steel and create a striking design. Three different strengths of steel were used, progressing from regular strength to high strength as the loads accumulate near the base reaction points. Another outstanding truss structure is the Greater New Orleans Bridge (Figure 4.17) with a central span of 1575 feet.

There are unlimited opportunities for utilizing the truss form for structures covering an area and carrying a distributed surface loading. The three-dimensional extension of the basic triangular form is the tetrahedron; systems of tetrahedrons can be combined to form any desired shape. The assemblages of tetrahedrons are called space trusses or pin-jointed space frames. The space truss is a favorite form for building a large, light, and stiff structure such as that required in a radar antenna or a radio telescope (Figure 1.7*h*).

**Fig. 4.17**
Greater New Orleans Bridge.

*Bethlehem Steel*

## 4.3 STRUCTURES TRANSMITTING LOADS BY BENDING ACTION

The problem of supporting a single load $Q$ at the center of span $AB$ is again considered. Suppose we are faced with supporting the load at a vertical position near the elevation of the supports, and that user space requirements necessitate as much clear space as possible both above and below a line through points $A$ and $B$ (Figure 4.18). Tension or compression structures are not feasible because the horizontal forces grow rapidly as the height of the structure is decreased. A shallow truss might be considered as a possible solution, but we must be aware of what a decrease in truss depth does to the forces in the truss members. A truss with $n$ panels and diagonal members at 45° slope is shown in Figure 4.19. We determine the magnitude of force in each member by starting at the right reaction and considering the equilibrium of each joint across the truss. An inspec-

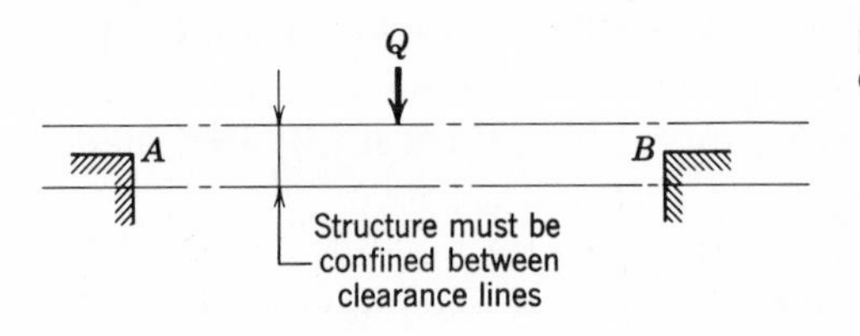


**Fig. 4.18**
Constrained loading situation.

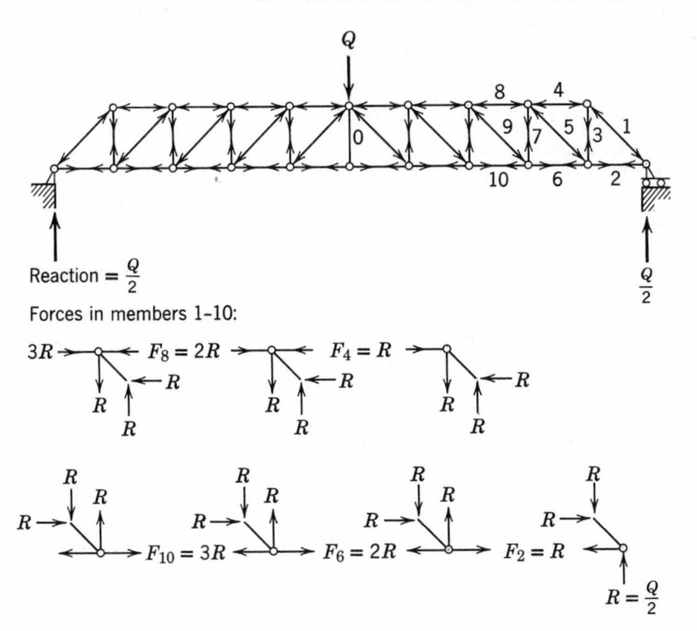


**Fig. 4.19**
Forces in a truss structure.

tion of the member forces in Figure 4.19 reveals that the forces in the vertical and diagonal members of the truss are independent of truss depth, depending only on the angle of inclination of the diagonal. However, with 45° sloping diagonals, the force in the horizontal chord members increases by $R$ every time we cross a joint (working from the reaction to the center of the truss).

Thus the magnitude of the force in the chord increases linearly with decreasing height of truss, with the top chord in compression and the bottom in tension. Furthermore, the chord forces are maximum at the center of the truss and minimum at the ends. The flow of force in a simple truss is always easily determined by elementary equilibrium considerations.

Although a very shallow truss is a feasible solution for this situation, there will be a large number of joints and individual members in the truss, and the cross-sectional area of each chord member will become large. The truss can be replaced by a solid element, called a *beam* (Figure 4.20), in which the top portion of the beam acts as a compression chord, and the bottom portion as the tension chord. In addition to resisting the tension and compression forces, the beam fibers also must carry the forces that would have been resisted by the truss verticals and diagonals. The transmission of load perpendicular to the axis of a long member is called *beam*

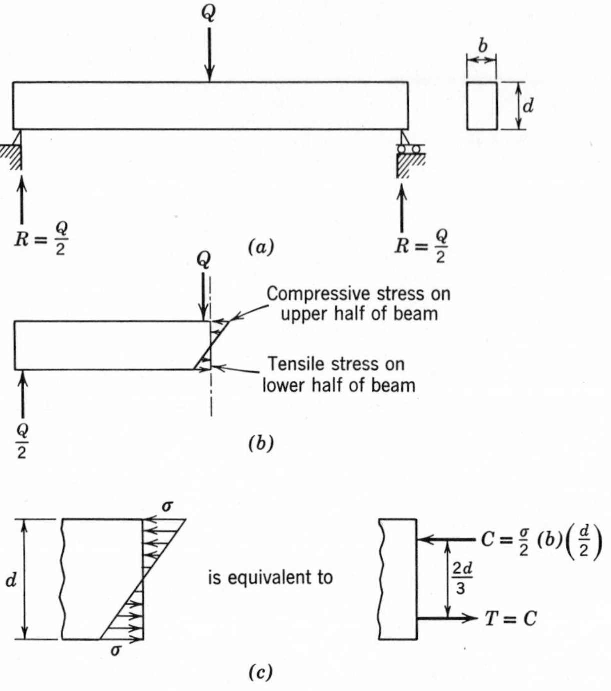


**Fig. 4.20**
Bending structure. (*a*) Simple beam. (*b*) Bending action. (*c*) Resultants of bending stress.

*action.* The redirected perpendicular force is transmitted to the ends of the beam span by a complex state of stress; the axial fiber stresses (analogous to chord member forces in a truss) are called *bending stresses,* while the transverse stresses (analogous to web member forces in a truss) are called *shearing stresses.* The continuous, single element called the beam is one of our most important structural forms.

Many load-carrying members found in nature are proportioned such that they are able to resist bending moment. The limb of a tree is always subjected to bending caused by the weight of the limb and its leaves; it is a cantilever with a tapered cross section because of the decreasing bending moment as the end of the limb is approached. Bones in mammals are often overloaded in bending, leading to failures that we customarily call fractures. Plants growing vertically, including trees, are subjected to bending moment every time the wind blows. In addition to carrying bending, vertical elements in nature are loaded in compression by gravity. Their characteristic round shape, often hollow as in a plant stem, is the optimum shape for resisting compression and bending moment applied in a randomly oriented plane. The so-called columns in a building are always subjected to both axial compression and bending, and are more appropriately called beam-columns.

Force transmission by bending is not efficient in comparison to axial force transmission. In Figure 4.21 we show a compression structure and a beam carrying the same load of 10 kips on a span of 24 ft. If the member sizes are 6 in. by 6 in. in both structures, the compression members in Figure 4.21*a* must resist a uniform stress of 361 psi over their entire lengths, while the beam in Figure 4.21*b* has zero bending stresses at its ends and a stress condition at midspan varying from 20,000 psi compression in the top fiber to 20,000 psi tension in the bottom fiber. The beam stress state is typical of a bending member, in which there is a large amount of material stressed to levels far below the maximum stress, because of both the variation of bending moment along the length of the beam and the distribution of stresses across the beam section. The beam

**Fig.4.21**
Comparison of compression and bending structure.

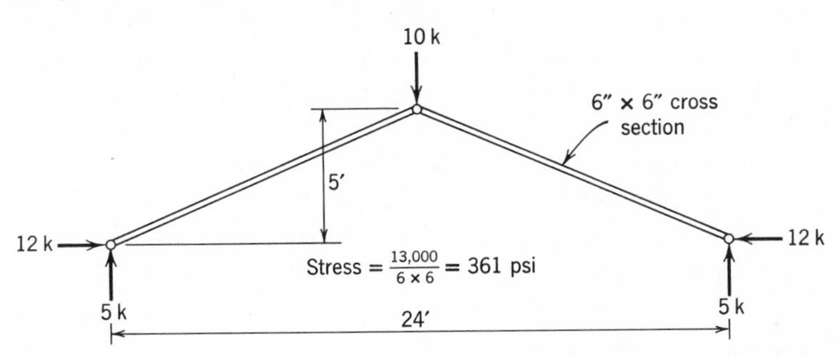


(*a*) Compression structure.

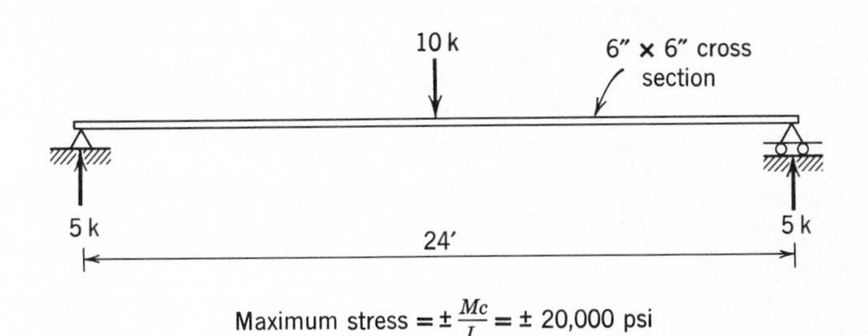


(*b*) Bending structure.

(*c*) Efficient beam shapes.

does have one major advantage in that it exerts no horizontal loads on its supports. The load of 10 kips is applied normal to the beam and leaves the beam in the same direction at the two supports.

Man has evolved beam shapes that partially compensate for the low stresses occurring in the central region of the cross section. These shapes are suitable only for those cases where loading is constrained to act in a unique plane; they have the characteristic of concentrating the material as far away from the center of the beam as possible (measured in the plane of the loading). The resulting heavy flanges (Figure 4.21*c*) are thus subjected to stress levels near the maximum extreme fiber stress state. The prestressed concrete beams designed for the Solleks River Bridge have this type of cross section. The cantilever steel beams for the roof of a large building (Figure 4.22) have a similar cross section that is tapered to account for increasing bending near the supporting tower. You should rationalize the unsuitability of this special shape for a natural structure (such as a stalk of wheat) which must be able to resist wind loading from any direction.

We have neglected the discussion of shear stresses and their effect on beam behavior. For now, it is sufficient to realize that the shear stresses are not ordinarily as serious as the bending stresses, and the two effects are relatively independent since maximum shear stress in a cross section usually occurs where bending stress is zero, and extreme fibers have zero shear stress and maximum bending stress.

Bending members place special requirements on materials since the material must be capable of carrying both tensile and compressive stresses

**Fig. 4.22**
Palace of Labour, Turin, Italy.

*PCA*

of about the same magnitude. This creates no problem with most metals, except for the possibility of buckling tendencies on the compression side of the beam, but it rules out the use of a material weak in tension, such as plain concrete and cast iron. The judicious combination of concrete and steel called reinforced concrete is widely used in structures acting in bending; compressive stresses are carried by the concrete and tensile stresses by the steel. Prestressed concrete also utilizes steel tendons embedded in the tensile side of the beam, but the stressing of the steel prior to loading produces compressive stresses that partially cancel the usual tensile stresses.

The placement and type of supporting reaction for a beam span are important design considerations that can be used to improve the efficiency of this structural form. The hinged-end beam in Figure 4.23*a* has a maximum bending moment of one-half the span times either reaction $= QL/4$. The moment is zero at both ends of the span. The beam must be designed to resist the maximum moment, and if we choose a cross section that is uniform, a substantial part of the beam operates at a low stress level.

Consider the same beam but with both ends fixed into rigid supports

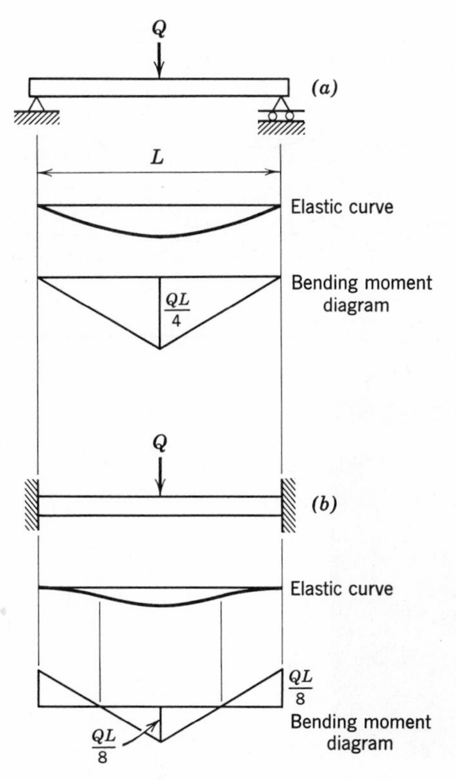


**Fig. 4.23**
Effect of support conditions on beam behavior. (*a*) Beam with hinged end (simply supported). (*b*) Beam with fixed ends.

(Figure 4.23*b*). The fixed ends prevent any rotation at the ends of the beam. The deflected shape (or elastic curve) and the moment diagram illustrate two crucial points. First, the moment diagram passes through zero at two points called inflection points; the curvature of the elastic curve changes sign at these locations. Second, the maximum moment is only half that of the simple span beam, with the peak value of $QL/8$ acting at three locations in the span. The disadvantage of this type of support condition is that massive, rigid, and usually expensive foundations are required to provide a high degree of end fixity.

The use of continuous spans over a series of supports enables the beam form to retain the advantages of the fixed end beam without having to fix the ends. The beams of the Solleks River Bridge are designed to act as a three-span, continuous beam for resisting the truck loading. As shown in Figure 4.24*a*, the interior spans of a beam continuous over five supports will have a distribution of bending moment very close to that found in a fixed-end beam. The continuous beam is used extensively in many types of

**Fig. 4.24**
Multispan beams.
(*a*) Behavior of four-span beam.

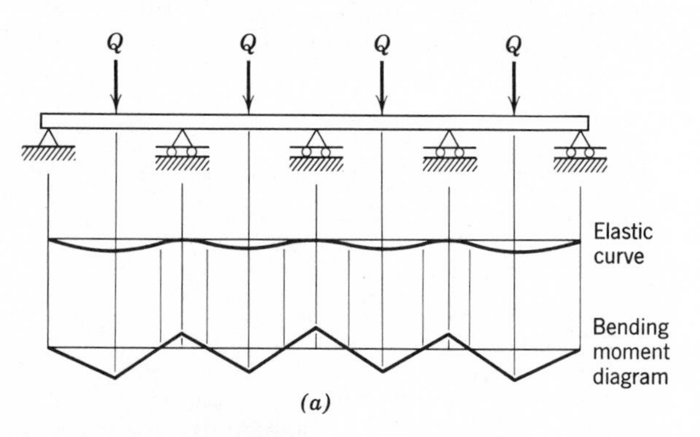

(*b*) San Mateo Creek Bridge, California.

AISC

construction. The multispan San Mateo Creek Bridge in Figure 4.24*b* was a winning design in the AISC 1970 Prize Bridge Competition. The jurors commented: "a graceful structure that seems to grow out of its surroundings. It combines form and grace to create a pleasant viewing experience. The details are well executed."

A variation of the basic beam form is the *frame,* made up of a number of beam elements. Returning to the problem of supporting a load $Q$ at the center of a span $L$, a frame structure suitable for this problem is illustrated in Figure 4.25. The joints at points $C$ and $D$ are fully continuous, while the support connections at $A$ and $B$ are shown as hinged in this example. The load flows through both sides of the frame, producing both vertical and horizontal reactions at each support point $A$ and $B$. All members of the structure carry both bending and axial load.

The ends of the horizontal member $CD$ (beam) are partially restrained

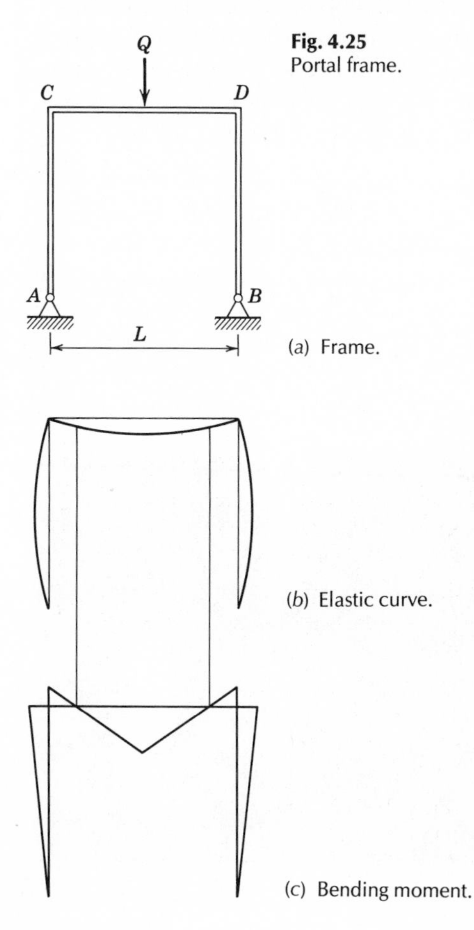


**Fig. 4.25**
Portal frame.

(*a*) Frame.

(*b*) Elastic curve.

(*c*) Bending moment.

from rotation by the vertical members $AC$ and $BD$ (called beam-columns). The deflected shape and moment diagram in Figure 4.25 can be sketched from an elementary knowledge of beam action, support details, and symmetry considerations. We should recognize that the frame can carry any type of in-plane loading applied at any point along the frame from points $A$ to $B$, including horizontal forces such as those produced by wind pressures acting on a building wall.

Another frame form for a single span is shown in Figure 4.26*a*; it is used widely in commercial and industrial buildings. The hockey rink in Figure 4.26*b* utilizes hinged laminated timber frames to support its roof. By combining a series of rectangular frames, we evolve the *multistory frame* shown in Figure 4.27*a*. The behavior of multistory frames varies considerably depending on the types of connections; one with rigid joints is capable of resisting horizontal wind loads strictly by bending of the members, while one with beams fastened to the columns with hinged connections must have additional structural members (such as the X-bracing in Figure 4.27*c*) to account for horizontal wind loads. A three-dimensional space may be enclosed by using frames in two perpendicular planes (i.e., in the plane of Figure 4.27 and normal to it). See Figures 1.7*b* and *i*.

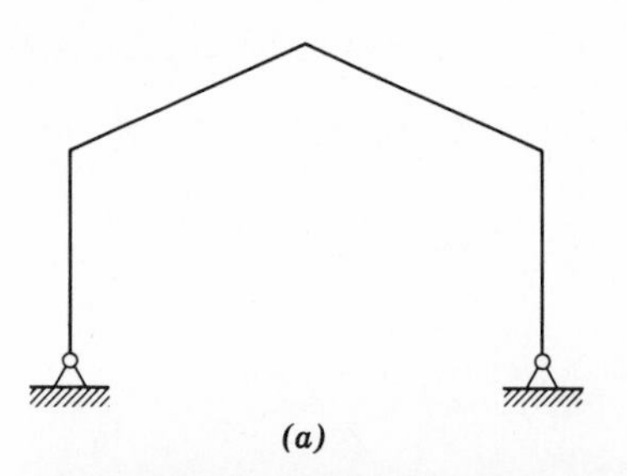

**Fig. 4.26**
Single span frames. (*a*) Gable frame. (*b*) Hall Hockey Rink, Hill School, Pottsdown, Pa.

AITC

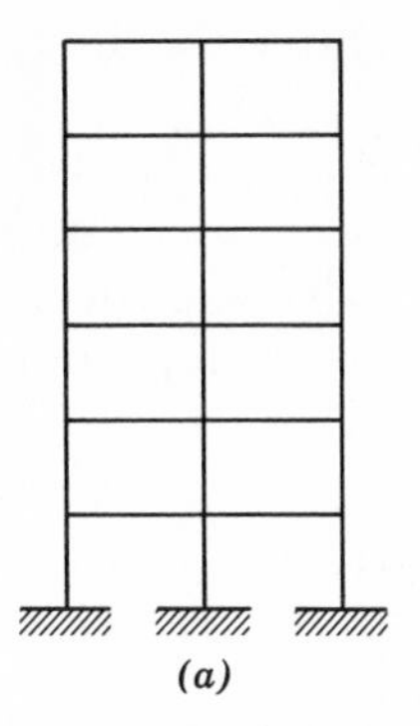
(a)

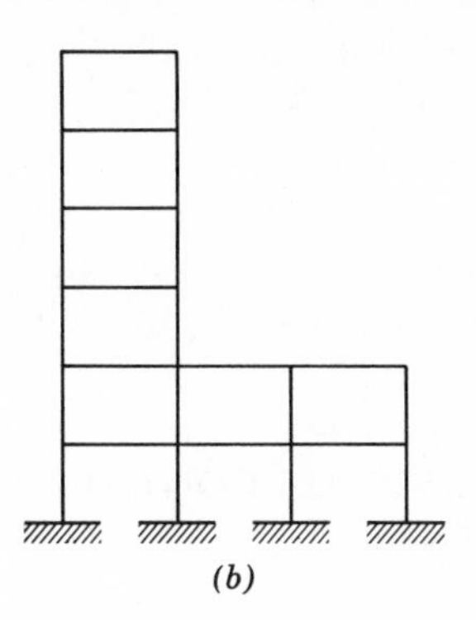
(b)

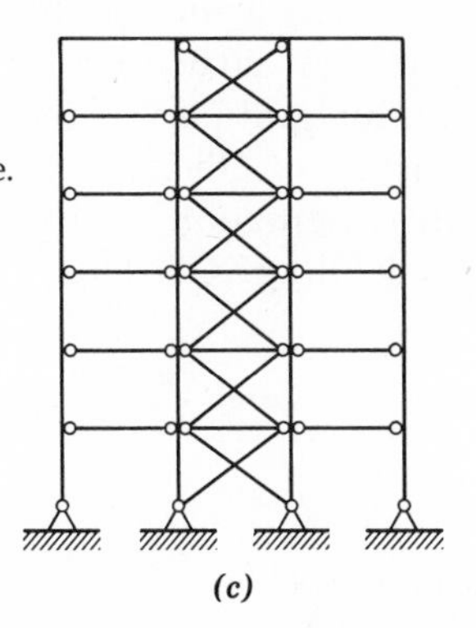
(c)

*First National Bank of Chicago*

**Fig. 4.27**
Multistory framed structures.
(*a*) Symmetrical rigid frame.
(*b*) Unsymmetrical rigid frame.
(*c*) Braced frame.
(*d*) First National Bank of Chicago building.

A sizable percentage of modern construction is dependent on the frame form, even though it is often less efficient structurally than other forms. The clear rectangular openings afforded by a conventional frame make up highly functional spaces needed for the organized activities of man.

The precise configuration of a multistory, three-dimensional frame is influenced strongly by the wind pressure loading on the building. The forces generated by wind loadings become increasingly critical as the height of the building is increased, and may control the design of many members in the structure. Some tall, slender skyscrapers are designed to act as giant cantilevered tubes. The horizontal wind loadings are transmitted to the foundation of the building by bending action, with one side of the building carrying the tensile force and the other the compressive force. Some are tapered to provide increased resistance to bending near the base (Figure 4.27*d*). The vertical members of such structures must carry both the gravity-induced compressive loadings as well as bending action induced by the weight of the structure and the wind-induced forces. We shall consider the behavior and design of high-rise building systems in more detail in Volumes 3 and 4.

Bending action is also the primary load-carrying mechanism in slab floor systems. A slab can be viewed as a continuum of beam elements, spanning in either one or two directions to supporting walls or girders. Reinforced concrete is used extensively in slabs for both bridges and buildings.

## 4.4 STRUCTURES COMPOSED OF CONTINUOUS FLAT AND CURVED ELEMENTS

Spaces in buildings or other enclosed structures are defined by flat or curved surfaces. Although certain types of surfaces (nonloadbearing partitions, glass curtain walls, etc.) do not participate in carrying loads, there are an increasing number of structures being built in which the space-defining surfaces are an integral part of the structural form.

The load-carrying mechanism and efficiency of a planar surface structural element depends on the orientation of the element with respect to the direction of the resisted loading. Consider the boxlike structure in

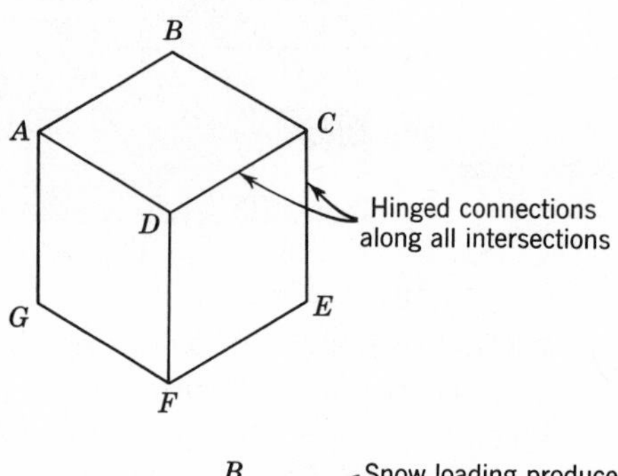


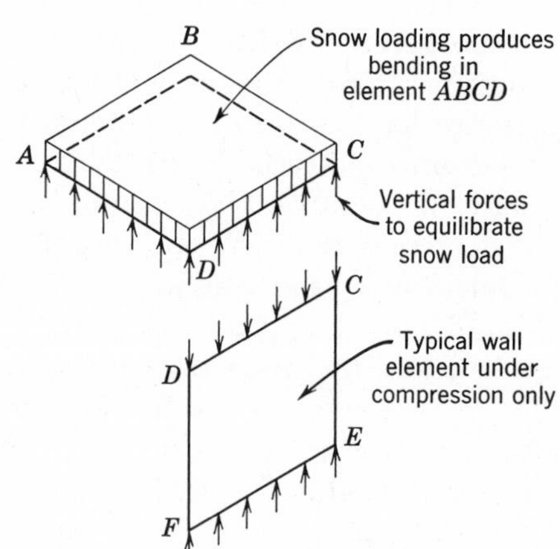


**Fig. 4.28**
Box structure subjected to snow load.

Figure 4.28, and assume that the elements are connected by continuous hinges. This basic form is being used in modular housing.

The element *ABCD* acts in bending when a uniform snow load is applied to the top of the structure. The load is redirected to the vertical sides *ADFG, CEFD*, etc., and these four elements will be stressed in compression only, transmitting the forces from *ABCD* directly to the foundations. We should note that the condition of pure compression in the side elements is possible only if the element connections are true hinges; any type of restraint in the connections would induce bending moments in the vertical wall elements.

If the same structure is loaded by horizontal wind forces instead of the snow loading, the vertical element facing the wind will redirect the loading to its edges by bending (Figure 4.29). The horizontal edge reactions on *ADFG* produce forces on the two adjacent side elements and on the top element *ABCD*. The edge forces on *ABCD* are equilibrated by concentrated shearing resultants (Figure 4.29*c*) and the entire element is in a complex state of in-plane stress. The side walls parallel to the wind direction

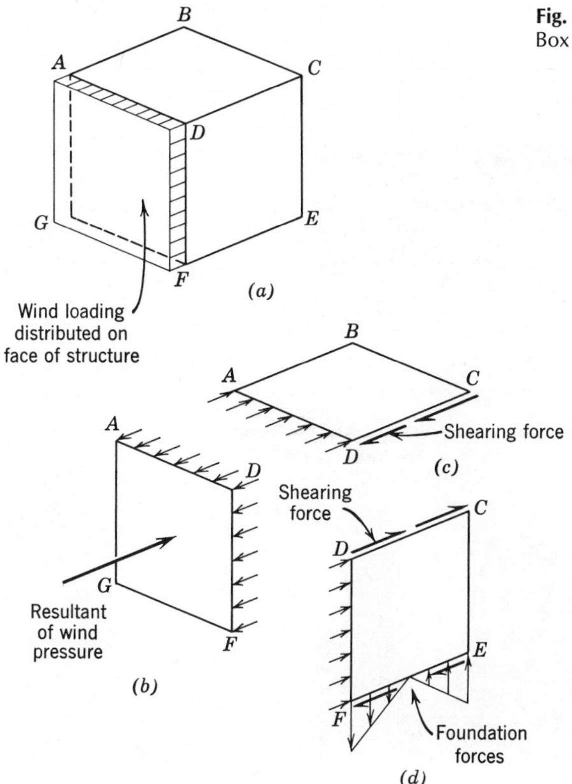


**Fig. 4.29**
Box structure subjected to wind load.

(Figure 4.29*d*) are in an even more complex state of in-plane stress since they must carry concentrated shear forces from *ABCD* and the edge forces from the wind-loaded element. This combination of element and loading is called *shear wall action* because most of the load is transmitted to the foundation by in-plane shearing stress.

This simple example illustrates the fact that two modes of force transmission are found in any rectangular surface structure: *bending and direct stress.* The bending is produced by loads acting normal to the surface, while the direct stress results from the in-plane forces. Most surfaces in

**Fig. 4.30**
Folded plate structures.

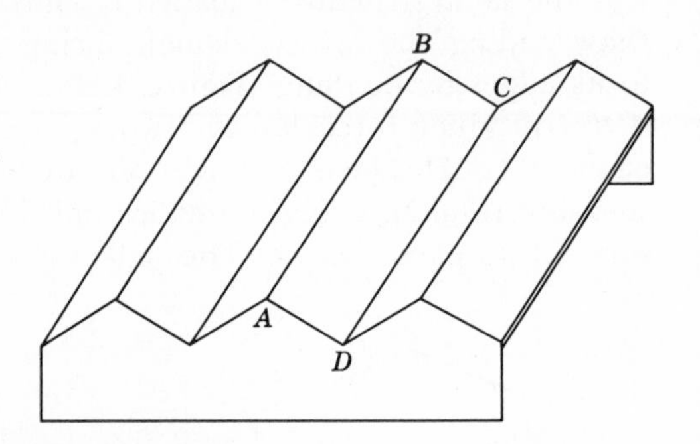


(*a*) Folded plate roof.

(*b*) Chapel for St. Joseph's Hospital, Burbank, Calif.

*PCA*

this type of structure must be able to act in either mode when subjected to the full range of possible loading conditions.

More efficient usage of material often can be made by arranging the flat elements to be at varying angles to the direction of loading. Typical *folded roof structures* are shown in Figure 4.30. An element such as *ABCD* must function in both bending and direct stress when the structure is loaded with snow; the bending action redirects the snow load forces to the edges *AB* and *CD*, while the direct stress action carries the loads imposed along edges *AB* and *CD* by the two adjacent flat elements. The same idea can be extended to produce folded, domelike structures such as the 400-ft diameter University of Illinois fieldhouse (Figure 4.31) and striking peaked buildings such as the U.S.A.F. Academy Chapel (Figure 4.32). The architectural and structural opportunities afforded by the folded plate form are essentially unlimited. A less noticeable but highly important usage of this structural form is in the deck (roadway) of many modern bridges; the orthotropic deck form is illustrated in Figure 4.33.

The search for the most efficient shape for transmitting a distributed loading leads to curved surface structures called *shells*. There are numerous types of doubly curved, thin shells occurring in nature, including the egg shell and a large number of shell-enclosed marine animals. Man has shown great interest in adapting the shell form to his own creations.

Consider the problem of carrying a snow load on a roof covering a

**Fig. 4.31**
University of Illinois fieldhouse. *PCA*

**Fig. 4.32**
U.S. Air Force Academy Chapel.

*U.S. Air Force*

circular building. If the uniform loading was along a line, we could use a parabolic arch stressed in compression only. For the uniform surface loading, a doubly curved *dome* can be used which will carry the snow load by direct stresses only (Figure 4.34). This dome shape can be created by revolving a parabolic or circular arch about a vertical axis through its highest point.

The most efficient use of the curved surface is in evolving shapes that have little or no bending and act mainly in compression or tension. Such structures are called *membrane shells*; the accompanying membrane stress states are relatively easy to determine in many types of shells. The absence of bending stresses makes the membrane shell highly efficient. In actual practice, we find that most membrane shell structures do not function purely in the membrane state mainly because of problems arising from

**Fig. 4.33**
Folded structure—orthotropic deck section, Poplar Street Bridge, St. Louis, Mo.

*Bethlehem Steel*

support conditions at their boundaries. The resulting localized bending stresses are difficult to calculate.

Man-made shells abound in many aspects of modern life. Large structures such as auditoriums, arenas, airline terminals, stores, and even restaurants and churches are roofed by shell structures of many varieties. The small sampling of structural shells shown in Figures 4.35 and 4.36 represents some of the many shells that have been built in the past 30 years. Airplane fuselages and wings (Figure 1.7*g*) are structural shells with internal beams acting as integral stiffeners; similar construction is used in building space rockets. Water and fuel storage tanks are shells stressed primarily in tension. Many dams (Figure 1.7*a*) are doubly curved shells, transferring water pressure into compressive thrusts against the

**Fig. 4.34**
Shell form (dome).

**Fig. 4.35** Shell structures.
(a) Palais des Expositions, Paris, 721-ft. span. *PCA*

(*b*) Priory Church and School, St. Louis, Mo. *PCA*

(c) TWA Terminal building, Kennedy International Airport *PCA*

**Fig. 4.36** Hyperbolic paraboloid shells.
(*a*) (above) Los Manantiales restaurant, Xochimilco, Mexico (by Candela).
*PCA*

(*b*) (at side) Scioto Downs grandstand, Columbus, Ohio.
*PCA*

(*c*) (below) Placing concrete, St. Edmund's Episcopal Church, Elm Grove, Wis.

*PCA*

rock foundations. Optimum shapes for dams are often obtained by loading flexible membrane models with fluid pressure, measuring the shape of the tensile structure, and then reversing the configuration for a compressive arch dam.

The majority of shell structures for buildings are built of reinforced concrete; the required curved surfaces are usually difficult to accomplish with other materials (such as steel) which cannot be molded into shape. Concrete placement for a large shell roof is shown in Figure 4.36*c*. The hyperbolic paraboloid shells in Figure 4.36 are made up of straight line generators, and can be made from reinforced concrete, timber or corrugated or folded sheets of light gage steel. A steel hyperbolic paraboloid form is shown in Figure 1.7*c*. Pressure vessel shells most often are made of steel and other metals, although there is an increasing usage of reinforced and prestressed concrete vessels for such applications as water storage tanks and containment vessels for nuclear power stations (Figure 1.7*j*).

The high efficiency and pleasing lines of a well-designed shell make it an attractive structural form. Unfortunately, there are two factors that tend to inhibit the usage of shells—difficulty in analysis and relatively high cost of construction.

## 4.5 COMBINED FORMS

A large number of structures utilize a combination of several basic forms. The Raleigh, North Carolina pavilion shown at the front of Chapter 6 combines the arch and the cable in an ingenious manner, in which the tensile reactions of the cable suspended roof are resisted by the two inclined arches. The water tank in Figure 1.7*e* has a tensile shell form for containing the water, with a single beam-column to support the tank and resist wind loading. Almost all structures have some bending elements in them; the floor system of a truss bridge, which transfers vehicle loads to the joints of the truss, is an indispensable part of the structure.

## 4.6 SUMMARY

Several basic structural forms have been studied from the standpoint of force transmission through space. We have classified these forms as follows.

1. *Tension or compression structures,* in which load is transmitted by a single state of stress. The pure tension structure is form-active in that it takes on a certain geometry for any given loading. It provides a highly efficient usage of material. The compression structure also can be quite efficient, but its load capacity is usually limited by buckling rather than by the inherent strength of the construction material.

2. Structures with tension and compression elements (*trusses*). The truss, made up of a pinned assemblage of members, carries variable loadings through the mechanism of varying magnitude of the member axial forces. It has excellent efficiency and is usually very stiff.
3. *Bending or flexural structures,* in which load normal to the member is transmitted along the member by bending action. The typical bending structure is not as efficient as a pure tension or compression structure, but it remains one of our most widely used forms because of its simplicity and adaptability to almost any situation.
4. *Surface structures,* acting mainly by stresses in a continuous surface. The surface form often combines behavioral aspects of the other simpler forms, and is accordingly the most difficult to analyze and design. Force transmission by in-plane or membrane stresses in properly shaped shells provides economical use of material. The shell form also affords the most opportunities for assuring high esthetic qualities in the structure.

You are urged to delve deeper into the topic of structural form by consulting the suggested reading at the end of the chapter, and by studying the innumerable structural forms that are encountered in everyday life, both natural and man-made. Continuing critical analysis of existing structures and those proposed on paper will help to develop your ability to answer three key questions: (a) How does the structure transmit its loads to its supporting foundation? (b) What type of deformations and stresses are produced by the force transmission? (c) How could the form be improved in structural efficiency without compromising its function and esthetic appeal? An increased appreciation of form will come with a better understanding of structural behavior.

We conclude this important topic by noting that routine analysis and design of many conventional structural systems can now be done almost entirely with the digital computer. This development is leading to a deemphasis of detailed hand calculations, and is helping to create a renewed demand for structural engineers who can devise new forms and choose the best structural system for any given situation.

## *Suggested Reading**

Angerer, Fred [1961]: *Surface Structure in Buildings,* Reinhold, New York.
Bill, Max [1969]: *Robert Maillart—Bridges and Construction,* Praeger, New York.

* This list is long because most libraries will have only some of the books; you should study several of those available in your library.

Engel, Heinrich [1967]: *Structure Systems,* Deutsche Verlags-Anstalt GmbH, Stuttgart.
Huxtable, Ada Louise [1960]: *Pier Luigi Nervi,* George Braziller, New York.
McGuire, William [1968]: *Steel Structures,* Prentice-Hall, Englewood Cliffs, N.J. (Chapter 2).
Museum of Modern Art [1964]: *Twentieth Century Engineering,* New York.
Salvadori, Mario, and Heller, Robert [1963]: *Structure in Architecture,* Prentice-Hall, Englewood Cliffs, N.J.
Torroja, Eduardo [1958a]: *Philosophy of Structures,* University of California Press, Berkeley.
Torroja, Eduardo [1958b]: *The Structures of Eduardo Torroja,* F.W. Dodge Corp., New York.

## PROBLEMS

4.1 Using a light chain or cord as a tension member, a piece of pegboard as a supporting base, and any convenient set of weights, study the funicular shapes for the loadings shown in Figure 4.4 (*a* and *b*). If the length of the chain is varied, do the relative values of the ordinates remain constant? Can you compute these shapes? If so, compare the measured and calculated shapes.

4.2 Sketch several alternate structural forms for the Solleks River Bridge and comment on their potential usefulness.

4.3 Maximum existing span lengths are about 1700 ft for arch bridges and 4000 ft for suspension bridges. Why is there a difference of more than 2 to 1 in these maximum spans achieved to date? Is there an upper limit on steel cable suspension bridges, or can we continue making larger and larger spans by merely increasing the size of the cables?

4.4 An auditorium convention center building is to have a clear span roof 200 ft on a side. Think about the types of structural forms suitable for this situation and sketch several alternatives. What is your main difficulty faced in this problem?

4.5 Select one of the prominent structural forms (i.e., tension structure, arch, simple span beam, continuous beam, truss, flat or folded surface structure, shell, etc.) as a topic for special study. Using the resources of your library, write an essay on the development of the form, on the types of materials most frequently used in the form, and cite two actual structures that utilize the form. One of these structures should be contemporary and the other of historical interest. Suggested library sources include the references at the end of this chapter, similar books on structures and history of engineering, and magazines such as *Engineering News-Record, Civil Engineering, Architectural Record,* and *Acier-Stahl-Steel.*

4.6 Study a picture or drawing of any major suspension bridge (George Washington, Verrazano Narrows, Golden Gate, Oakland Bay, Tagus River, Severn, etc.). What types of forces act on the towers? What prevents the bridge from undergoing major changes in geometry under heavy unsymmetrical traffic loadings? How are the forces at the ends of the cables resisted?

4.7 In the history of structural engineering, we find a few individuals who clearly stand out as original and daring conceivers and builders of structures. These names include John Roebling (Brooklyn Bridge), James Eads (Eads Bridge), Robert Maillart (arch bridges and new forms of concrete buildings), Pier Luigi Nervi (shell structures and development of ferro-cemento), Felix Candela (hyperbolic paraboloid shells), Eduardo Torroja (shells, bridges, and buildings), David Steinman (long span bridges), and Fritz Leonhardt (prestressed and cable structures). Select one of these individuals, or another well-known engineer, and spend some time studying his writings and achievements. Your instructor can help guide you into the literature.

4.8 Choose three examples from nature (e.g., trees, shells) illustrating the interaction of form and structural action. Discuss this interaction.

4.9 Discuss the relationship between the shape of the bodies of three types of animals as a function of the loading or environment they experience.

4.10 Cite three examples where parts of the human body are ideally formed to fulfill structural roles.

4.11 Discuss and compare various types of roofs appropriate for covering an area of 60 ft by 60 ft in terms of structural action, ease of construction, and other aspects.

4.12 Galileo, in his extraordinary book, *Dialogues Concerning Two New Sciences,* (Galileo [1638]) makes the following observation: "... it would be impossible to build up the bony structures of men, horses, or other animals so as to hold together and perform their normal functions if these animals were to be increased enormously in height; for this increase in height can be accomplished only by employing a material which is harder and stronger than usual, or by enlarging the size of the bones, thus changing their shape until the form and appearance of the animals suggest a monstrosity. . . . If the size of a body be diminished, the strength of that body is not diminished in the same proportion; indeed the smaller the body the greater its relative strength. Thus a small dog could probably carry on his back two or three dogs of his own size; but I believe that a horse could not carry even one of his own size."

Do you agree with Galileo? What implications does his statement have with regard to man-made structures? Think carefully about Galileo's insight and comment critically on it.

4.13 The stayed girder bridge of Figure 1.7*f* is a form that is being used increasingly for medium-span bridges. Trace the transfer of load on the bridge deck through the superstructure to the foundations, and identify the basic forms utilized in this bridge.

4.14 Thousands of structures are used in forming grade separations at the intersections of super-highways (interstate routes and turnpikes) and other highways. During your next auto trip, observe these structures and comment on the following aspects: type of structure (simple span versus continuous span, frame, etc.), number of spans, configuration of supporting piers, materials of construction, and esthetic appeal. Note the differences as you pass from one state to the next. Prepare a sketch showing the major components of one of the bridges, and discuss flow of load through the structure.

## LAB EXERCISES

L4.1 A small structure is required to support a load of 100 lb at the center of a 24-in. span without deflecting more than 1/2 in. The maximum depth of the structure may not exceed 4 in., where depth is defined as the vertical distance between the end support point and the highest point on the structure. Using heavy, laminated illustration board and a quick drying cement, design and construct a structure that is suitable for this situation.

The merit of the solution will be judged on the basis of total weight of the structure.

L4.2 Repeat Lab Exercise 1, using balsa wood and cement for the construction.

L4.3 A surface structure is needed to support a working load of 30 psf on a 24 by 24 in. area. Balsa wood and a quick drying cement are the materials of construction. Working in teams of four members, propose at least three structures for this situation. After consultation with your instructor, build and test one structure. Performance requirements include a limitation of 1/4-in. maximum deflection under working load, and a minimum ultimate capacity of 75 psf. The loading is best performed by supporting the roof structure on a square steel framework and loading with cans of lead shot or steel punchings. Loads should be transmitted to the roof surface through rubber pads that are fastened to the suspended weights by small wires passing through the surface.

CHAPTER 5

# Introduction to Structural Analysis

Port Mann Bridge, near Vancouver, Canada. *Bethlehem Steel*

# 5

# Introduction to Structural Analysis

The design process, as discussed in Chapter 1, includes the determination of forces and displacements in both preliminary and final designs. This phase of the design effort is called *structural analysis;* it is normally the major quantitative part of design. There are many aspects to analysis. We construct an idealized mathematical model, impose loads and other environmental effects, and then predict the resulting performance through all stages of loading. In some structures the mathematical analysis is supplemented by physical analysis on reduced size models of the prototype structure. In any case, the end goal of structural analysis is to develop the necessary appreciation of behavior and to compare expected performance with the stated requirements. We must recognize that a high level of proficiency in analysis by itself can never produce a design.

The most fundamental form of analysis involves statically determinate structures, which is the topic of Chapter 6. In the present chapter, we shall explore the various assumptions and conditions upon which analysis is based. Such terms as *elastic behavior, linearity, superposition, statically determinate* and *statically indeterminate* structures, and *geometric stability* are an essential part of the vocabulary of structural analysis. We must develop an understanding of the meaning of these terms prior to embarking on analysis. However, a full appreciation of these concepts depends on a knowledge of the fundamentals of analysis; thus you are urged to return to this chapter after completing a study of Chapter 6.

## 5.1 ELASTIC AND LINEAR BEHAVIOR

Although much progress in mechanics can be made under the assumption that materials are *rigid,* we soon encounter problems in which the deformation of material under stress affects our results. Galileo Galilei, in the early part of the 17th century, faced such a problem. He attempted to analyze the bending strength of a cantilever beam but had no knowledge of the deformation of materials. His published work (Galileo [1638]) included his analysis of this problem. Figure 5.1 is his diagram of the problem and its solution, based on the principles of statics and on the assumption that the beam rotates around point $B$ at the bottom edge. Galileo's statics was correct, but the implied distribution of stresses was such that he underestimated the maximum stress by a factor of three. It was half a century later that Hooke, an English physicist, published his *De Potentia Restituva,* in which he states (Todhunter and Pearson [1886]):

> . . . the Rule or Law of Nature in every springing body is, that the force or power thereof to restore itself to its natural position is always proportionate to the distance or space it is removed therefrom . . . Nor is it observable in these bodies only, but in all other springy bodies whatsoever, whether metal, wood, stones, baked earths, hair, horns, silk, bones, sinews, glass and the like . . .

Based on Hooke's principle of the proportionality of force and deformation, Mariotte assumed the correct stress distribution in Galileo's beam and (aside from a minor error) obtained the solution to the bending problem about 200 years ago.

Hooke's major contribution was the observation that deformation is proportional to applied load. We now know that the actual relationship between stress and strain may differ from the simple proportionality that Hooke observed, but we do find that for most engineering materials a linear relationship holds with a good degree of accuracy at least for low stresses. For example, a typical carbon steel tension specimen exhibits the stress-strain curve shown in Figure 2.2. The plot is a straight line right up to the point where yielding commences. Clearly, Hooke's Law is applicable here. Because this behavior is simple to analyze and to model mathematically, and because it provides an excellent approximation for most materials in the usual range of stresses, we often assume for purposes of analysis that our material follows Hooke's Law, and we term the resulting behavior *linear.* The constant of proportionality (slope of the stress-strain curve) is called *Young's Modulus* or the *modulus of elasticity,* and is generally given the symbol $E$.

A linear material is a special case of the more general condition termed *elastic* behavior. A material is elastic when its stress-strain curve on unloading follows the loading curve. In the linear case, this is automatically satisfied since unloading is usually linear, but common materials such as rubber are nonlinear while still elastic.

Dover Publications

(a)

**Fig. 5.1**
Galileo's cantilever beam.

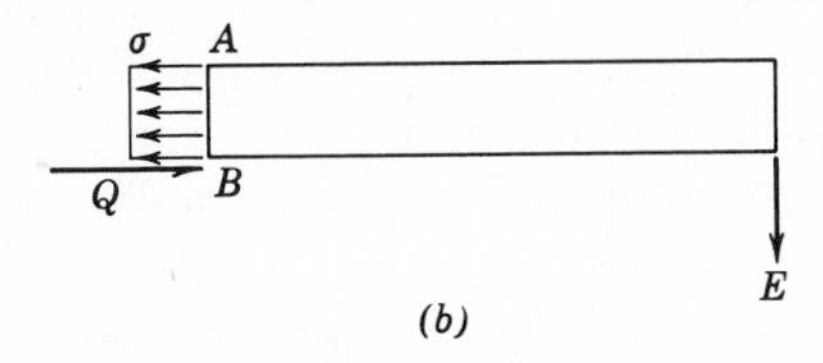


(b)

The concrete stress-strain curve of Figure 2.3 indicates an appreciable departure from linearity at about $0.50f_c'$. Beyond this stress, then, concrete is a nonlinear material. Because the assumption of linearity may be quite erroneous in many materials and structures, especially when the stresses are large, the engineer must be aware of such an assumption.

We may generalize the linearity assumption to an entire structure. When the displacements in a system of structural components are linear functions of the applied load or stress, then we have a *linear structure,* or a structure exhibiting *linear behavior*. Such structures are much simpler to analyze than nonlinear ones, so it is extremely important for us to learn how to identify such action. Nonlinear structures are encountered often enough, however, to justify a caution to the reader that many structural failures involve nonlinear behavior (e.g., buckling) and that the assumption of linearity must be made carefully and with full assurance that the structure will comply.

The identification of linear structural behavior is based on two conditions which, when taken together, form the sufficient condition for linearity. The first of these is that the materials be linear, and therefore elastic. Inelastic materials will always lead to nonlinear behavior. The second condition, more difficult to verify, is that the displacements of all parts of the structure under the applied loads be *small*. This raises the interesting question, "How small is small enough?" Most difficulties in recognition of nonlinear structures stem from difficulties in answering this question. Displacements are "small" when the equations of equilibrium, written in terms of the unloaded geometry, give results equal to those obtained from the same equilibrium equations written in terms of the final displaced shape. In other words, the initial and final geometries are the same for practical purposes. Skill in recognizing this condition will come by considering the question of small displacements in each of many examples and problems encountered in the future. A study of inelastic and nonlinear behavior is included in Volume 2, Chapter 17. The discussion there covers nonlinearity due to materials and geometric effects.

One of the most noteworthy types of nonlinear behavior is that encountered in beam-columns. If the initially kinked column of Figure 5.2*a* is loaded in compression as shown, it deflects into the shape indicated in Figure 5.2*b*. A free body diagram of the upper half of the member is drawn in Figure 5.2*c*, and the bending moment at midheight of the column is determined as $M = P\ (\epsilon + \delta)$. This moment depends on the

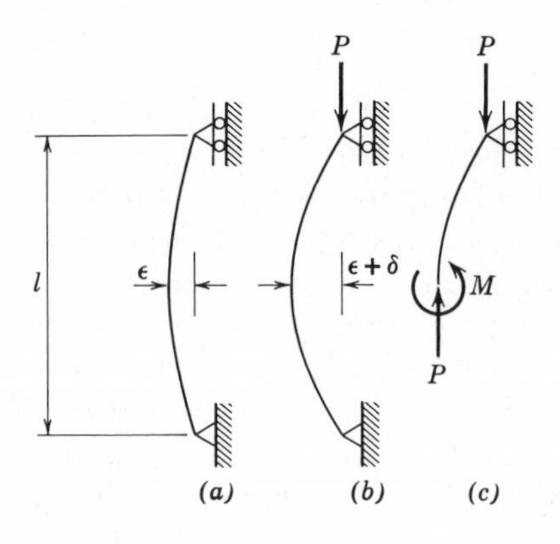


**Fig. 5.2**
Axially loaded column with initial kink. (*a*) Unloaded column. (*b*) Loaded column (*c*) Free body diagram of upper half.

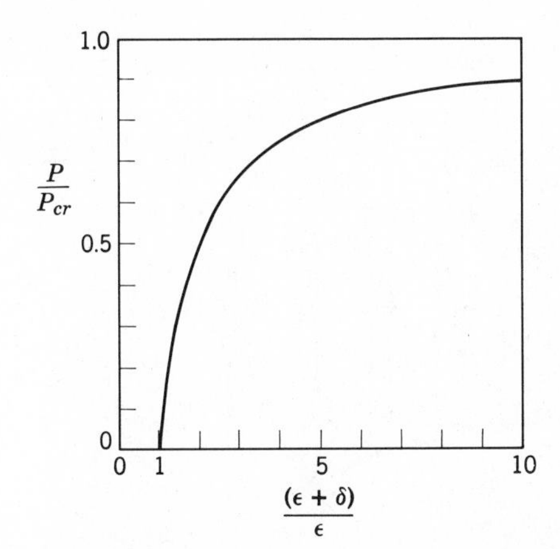


**Fig. 5.3**
Load-displacement curve for an imperfect column.

lateral deflection $\delta$. Since $\delta$ is also a function of $P$, the actual moment becomes a nonlinear function of the applied load $P$, and therefore the structure is nonlinear. This nonlinearity is a function of the slenderness of the column, since in a very stocky column, $\delta$ is not strongly dependent upon $P$. A plot of load versus midspan displacement for such an imperfect column appears in Figure 5.3. The term $P/P_{cr}$ is the ratio of the applied axial load $P$ to the Euler buckling load for a perfect column, $P_{cr}$.

One of the most tragic bridge failures in history involved the nonlinear performance of compression members. The Quebec Bridge collapsed during construction in 1907, taking the lives of about 75 workmen. Prior to the collapse, a compression chord member had shown a growing lateral displacement as its load increased. Engineers on the site noticed and measured the movement over a period of weeks. The lateral displacement in the unloaded position due to workmanship errors (and a construction yard accident resulting in the member in question being dropped and subsequently repaired) was similar to that noted as $\epsilon$ in Figure 5.2*a*. As the lateral displacement grew larger, the moment produced by the increasing axial load grew larger at an increasing rate. Although the engineers at the site became alarmed, they did not stop the work in time. When the member finally failed, the entire structure (anchor and cantilever arms of a cantilever truss) collapsed. Figure 5.4 shows the structure before and after the accident. The detailed report of this failure (Holgate, Kerry, and Colbraith [1908]) is worthy of careful study, for many of its lessons are valid today. As in most failures, many factors were involved. A major cause, however, was a lack of appreciation of the behavior of compression members.

**Fig. 5.4**
Failure of the Quebec Bridge, 1908.

*Canadian Government Board of Engineers*

***Example 5.1 Nonlinear Analysis*** Nonlinearity of the load-displacement relationship as a result of geometric effects is well illustrated by the bar structure of Figure 5.5*a*. We shall first examine case I, shown in Figure 5.5*a,* in which $s$ is small compared with $l$. This structure will be seen to be nonlinear. A subsequent examination of case II, shown in Figure 5.5*d,* will allow comparison of nonlinear versus linear behavior. The bars are pin-ended so that only axial forces are carried in the two members.

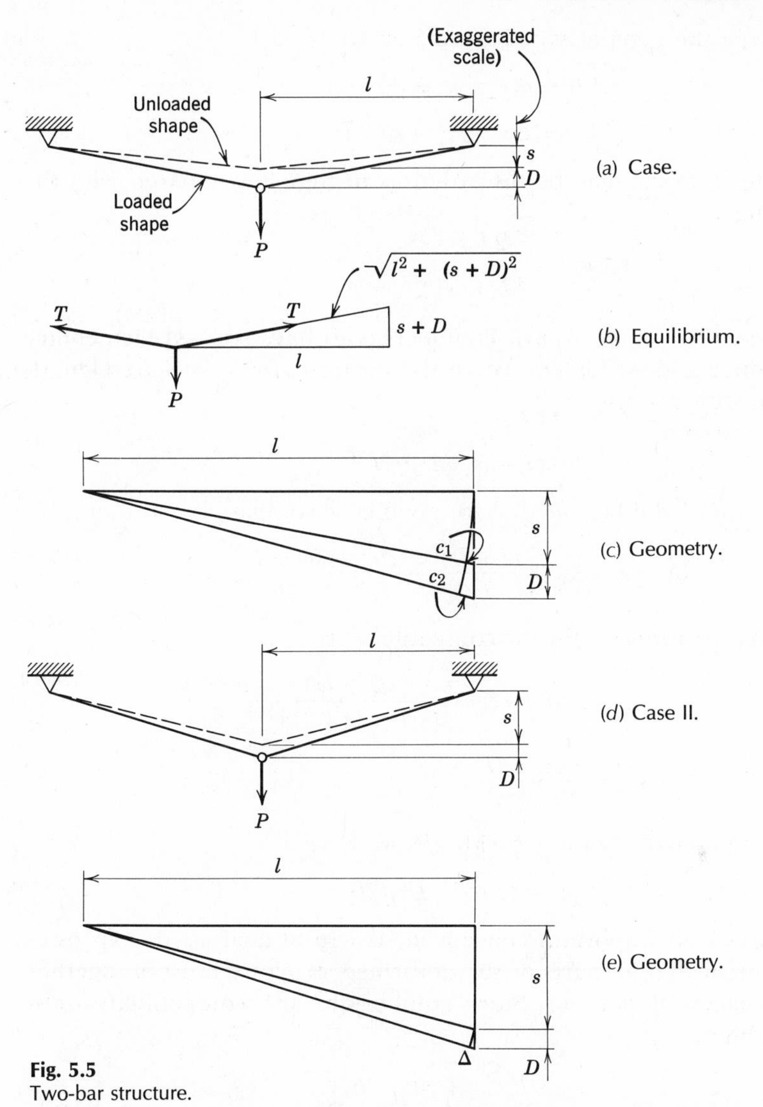


**Fig. 5.5**
Two-bar structure.

When the load $P$ is applied to the structure of Figure 5.5*a*, it displaces a distance $D$. The tensile force $T$ in the bars is found by equilibrium (Figure 5.5*b*) to be

$$T = P\sqrt{l^2 + (s + D)^2} \Big/ 2(s + D) \approx Pl/2(s + D)$$

The corresponding tensile strain required for equilibrium is

$$\epsilon_{Eq} = T/EA = Pl/\ [2EA(s + D)]$$

Considering the geometry of Figure 5.5*c*, we have

$$(l + c_1)^2 = s^2 + l^2$$

$$l^2 + 2lc_1 + c_1^2 = s^2 + l^2$$

in which the term $c_1^2$ may be discarded as negligible compared with the others, giving

$$2\,lc_1 = s^2$$

$$c_1 = s^2/2l$$

which is a familiar term to civil engineers who have studied elementary surveying, since it gives a length correction factor $c_1$ for a measured length over a slope with rise $s$.

Similarly, we find

$$c_2 = (s + D)^2/2l$$

The elongation of the bar needed for geometric compatibility is therefore

$$\Delta = c_2 - c_1 = \frac{(s + D)^2 - s^2}{2l} = \frac{2\,sD + D^2}{2l}$$

and the corresponding strain for compatibility is

$$\epsilon_c = \Delta/(l + c_1) = \frac{2\,sD + D^2}{2\,l[l + (s^2/2l)]}$$

$$= \frac{2\,sD + D^2}{2\,l^2 + s^2}$$

Neglecting $s^2$ as small compared with $2l^2$, we have

$$\epsilon_c = (2\,sD + D^2)/2l^2$$

Compatibility is an important concept in structural analysis. It expresses the requirement that all parts of the deformed structure must fit together during all stages of loading. Since equilibrium and compatibility must both be satisfied,

$$\epsilon_{Eq} = \epsilon_c$$

$$\frac{Pl}{2\,EA\,(s + D)} = \frac{2\,sD + D^2}{2\,l^2}$$

$$P = \frac{EA\,(s + D)\,(2\,sD + D^2)}{l^3}$$

$$P = (D^3 + 3\,sD^2 + 2\,s^2D)\,\frac{EA}{l^3} \tag{5.1}$$

$P$ is therefore a nonlinear function of the displacement $D$.

Now consider the structure of Figure 5.5*d*, in which $s$ is not small compared with $l$. Figure 5.5*b* again portrays the equilibrium condition

$$T = P\sqrt{l^2 + (s+D)^2} \Big/ 2\,(s+D) \approx P\sqrt{l^2 + s^2} \Big/ 2\,s$$

$$\epsilon_{Eq} = P\sqrt{l^2 + s^2} \Big/ 2\,EA\,s$$

The geometry, indicated in Figure 5.5*e*, gives (by similar triangles) the elongation $\Delta$

$$\frac{\Delta}{D} = \frac{s}{\sqrt{l^2 + s^2}}$$

and the strain

$$\epsilon_c = \frac{\Delta}{\sqrt{l^2 + s^2}} = \frac{Ds}{l^2 + s^2}$$

As before, we set $\epsilon_{Eq}$ equal to $\epsilon_c$ in order to satisfy equilibrium and compatibility

$$P\sqrt{l^2 + s^2} \Big/ 2\,EAs = Ds \Big/ (l^2 + s^2)$$

$$P = 2\,DEA\,s^2 \Big/ (l^2 + s^2)^{3/2} \tag{5.2}$$

in which $P$ is a linear function of the displacement $D$, indicating that the structure is linear. Figure 5.6 is a plot of the load-displacement relationship for cases I and II, with $EA = 30 \times 10^3$ kips, $l = 100$ in., $s_I$(case I) = 4 in., and $s_{II}$(case II) = 30 in. It is interesting to compare the nonlinear behavior of case I with that of the column (Figure 5.3).

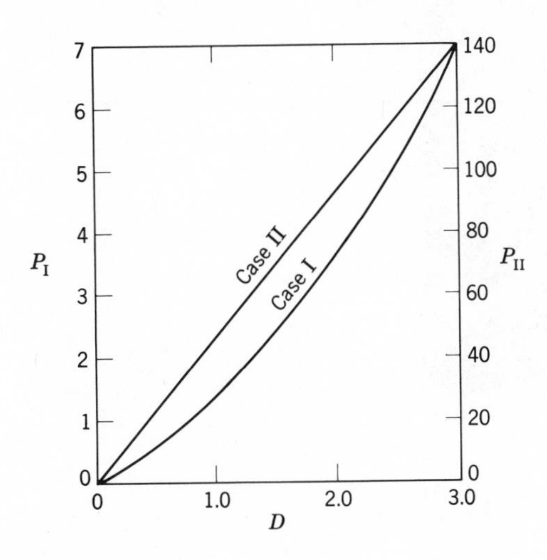


**Fig. 5.6**
Load-displacement curve for the two-bar structure.

The preceding example involved structures of the same basic form, but the analysis in each case involved somewhat different simplifications in the mathematics, with a resulting difference in the form of the relationship between load and deflection. In a sense, all structures are nonlinear, for a completely exact analysis would involve consideration of displacements in the equilibrium equations, without neglecting some terms because they are "small." Our analysis of structures would be most difficult, and no more accurate if we took this course. The engineer's task is to understand the behavior of his structure. He should avoid masking his analysis with considerations that might affect the accuracy only a fraction of a percentage. Members carrying compressive force are inherently nonlinear. Most other structures are linear; notable exceptions are cable structures such as cable roofs and bridges, and longspan arches. An interesting cable structure is shown in Figure 5.7. This is one of the world's largest pipeline bridges, crossing the Ohio River near Portsmouth. The nonlinear analysis of cable networks is discussed in Volume 2, Chapter 17.

## 5.2 THE PRINCIPLE OF SUPERPOSITION

The major reason for our interest in the assumption of linearity of structural behavior is that it allows us to use the *principle of superposition.* This principle means that the displacements resulting from each of a

**Fig. 5.7**
Pipeline Bridge over the Ohio River.

*Bethlehem Steel*

number of forces may be added to give the displacement resulting from the sum of the forces. Superposition also implies the converse: the forces corresponding to a number of displacements may be added to yield the force corresponding to the sum of the displacements.

As an example, consider the column of Figure 5.8. The stress at the lower end is $(P_1+P_2+P_3)/A$, which can be factored into $P_1/A+P_2/A+P_3/A$. The latter form shows that the total stress can be found as the sum of the stresses for each of the three loading conditions: the loads $P_1$, $P_2$, and $P_3$ applied separately. This would be the case even if the materials had been inelastic. The total shortening of the column, assuming that Hooke's law holds, is

$$(P_1+P_2+P_3)\,l_1/AE + (P_2+P_3)\,(l_2-l_1)/AE + P_3\,(l_3-l_2)/AE \tag{5.3}$$

This equation reduces to $P_1l_1/AE + P_2l_2/AE + P_3l_3/AE$, which again is seen to be the sum of the separate effects of each load. It is important to note that this useful result would not occur if the strain was not a linear function of the applied load.

Superposition, then, allows us to separate the loadings in any desired way, analyze the separate cases, and find the result for the sum of the load cases by adding the individual results. Superposition applies equally to forces, stresses, strains, and displacements.

As an example of the difficulty when the structure is nonlinear, consider

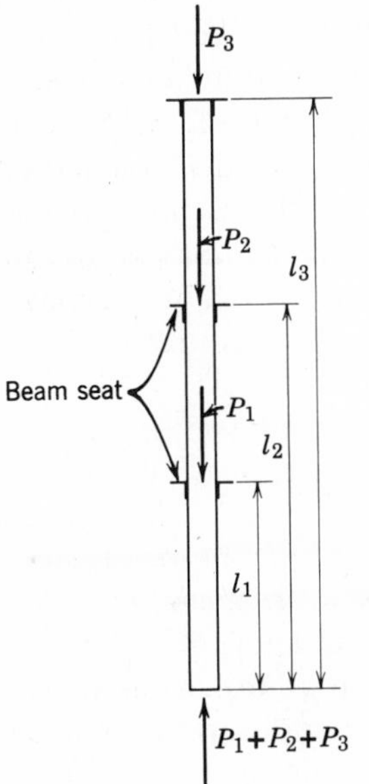


**Fig. 5.8**
Superposition of forces on a column.

Unadilla Laminated Products

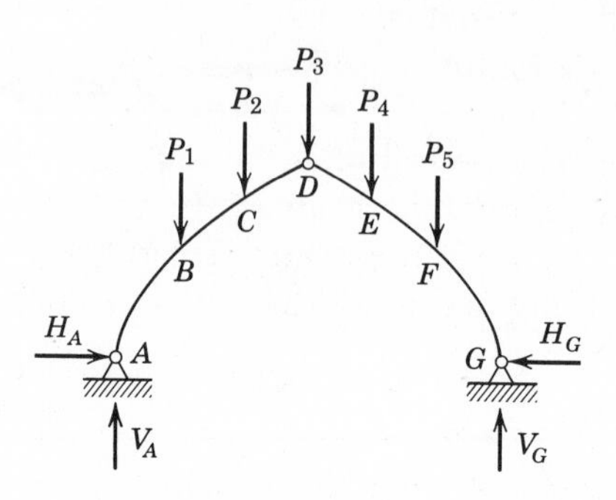


**Fig. 5.9**
Loaded three-hinged gable frame.

the behavior of the bar structure (case I) of Example 5.1, with the load-displacement relationship shown in Figure 5.6. The displacement for a load of 2 kips is 1.3 in. The displacement resulting from a load of 4 kips is only 2.2 in., not the result obtained by doubling the value for a 2 kip load. The principle of superposition must not be used for analysis of nonlinear structures.

The usefulness of the principle of superposition is illustrated by the three-hinged gable frame of Figure 5.9. In this case the reactions at the base hinges are required for the loading shown: loads $P_1$, $P_2$, etc. applied at equal horizontal intervals. The analysis of a three-hinged frame for a single concentrated load is much simpler than for a number of loads acting simultaneously. Section 6.4 indicates a graphical method for such an analysis. We may perform an analysis for a single load $P$ applied at point $B$, repeating the analysis for a single load at point $C$, and then at point $D$. The solution for a single load at point $E$ and at $F$ may be found by symmetry from the solutions for $C$ and $B$, respectively. The reactions for the original problem may be found by adding those found for each of the five cases.

## 5.3 DETERMINACY AND STABILITY

A structure may be classed as either statically determinate or indeterminate. In either case, the structure must be geometrically stable. In order to make the necessary classifications, the degree of indeterminacy is first investigated. The results of this investigation are conditional on subsequent verification of geometric stability. The following discussion may be supplemented by consulting the suggested reading material.

## Determinacy

A major aspect of structural analysis, covered extensively in Chapter 6 and in Volume 2, is the determination of the forces, moments, and torques (called forces for brevity) acting everywhere in the structure. A *statically determinate* structure is one for which the internal and external forces can be determined for any loading condition by resorting only to the static equations of equilibrium:

$$\begin{aligned} &\Sigma F_x = 0, \quad \Sigma F_y = 0, \quad \Sigma M_z = 0 \\ &\Sigma F_z = 0, \quad \Sigma M_x = 0, \quad \Sigma M_y = 0 \end{aligned} \tag{5.4}$$

For the purpose of clarity, most of our discussion will be in terms of two-dimensional structures, in which case only the first three equations can be used. You should be able to generalize to the three-dimensional case after some experience with two-dimensional structures. We note that determinacy does not depend on loading. Recognition of statically determinate structures is important because of the relative ease with which such structures can be analyzed.

A *statically indeterminate* structure is one for which equations in addition to those of static equilibrium are necessary to determine the forces. The additional equations usually involve the compatibility of deformations in the structure. When $n$ such additional equations are needed, the structure is said to be statically indeterminate to degree $n$.

In order to illustrate the concept of determinacy, consider the gable frame of Figure 5.9. To analyze the structure for the loading indicated on the diagram, we may draw a free body diagram of the left half (Figure 5.10*a*). There are four unknown forces, and only three equations of

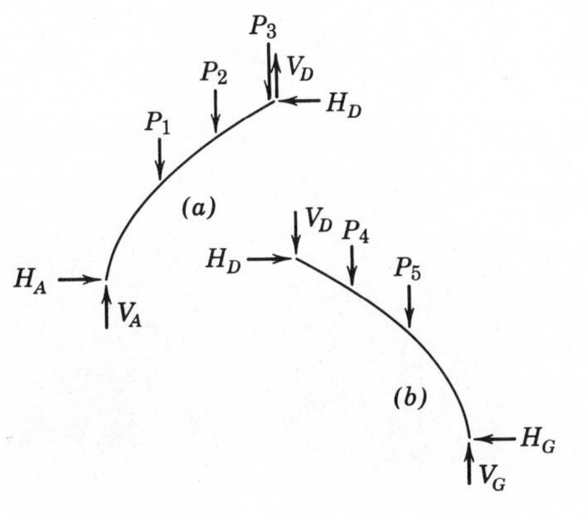


**Fig. 5.10**
Free body diagrams: gable frame.

statics available, preventing a determination of the four forces. The free body diagram of the right half (Figure 5.10*b*) also has four unknowns, but two of these are common to both free body diagrams. There is a total then of six unknown forces. Three equations of equilibrium can be written for each free body, yielding six equations for the six unknowns. The structure is therefore statically determinate.

The rigid frame of Figure 5.11 provides a second example. If the frame is cut at $B$, we introduce three unknown forces that must be computed: axial force, shear, and bending moment. When combined with the four reactions at the base, there are seven unknowns. With only six equations of equilibrium, the structure is statically indeterminate to the first degree since the number of unknowns exceeds the number of equilibrium equations by one. The cut at the top was not necessary in classifying the structure. A free body of the whole structure shows that there are four unknown forces and three equations of equilibrium, yielding the same result.

There are many methods for determination of the degree of static indeterminacy. Only one of these, general enough to handle all cases, is considered here. It makes use of the basic definition of static indeterminacy: we first set out to determine the total number of unknown *forces*

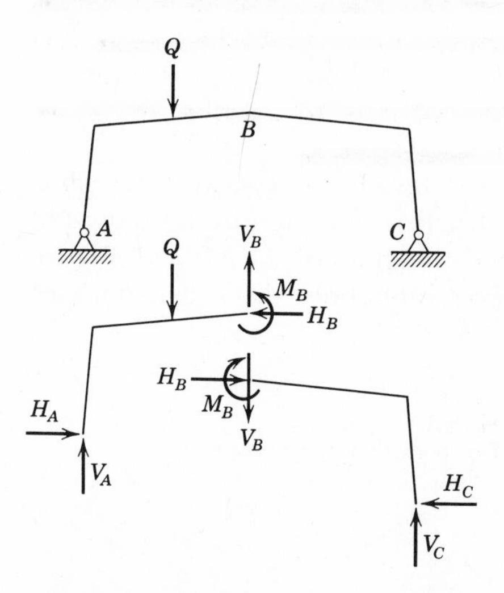


**Fig. 5.11**
Two-hinged rigid frame.

*Garceau Steel Structures*

in the structure, that is, the total number of axial, shear, and bending moment quantities necessary to completely determine the forces and moments everywhere in the structure. We then determine the number of available equations of statics. The degree of static indeterminacy $n$ is the difference between these two quantities. If $n = 0$, the structure is statically determinate.

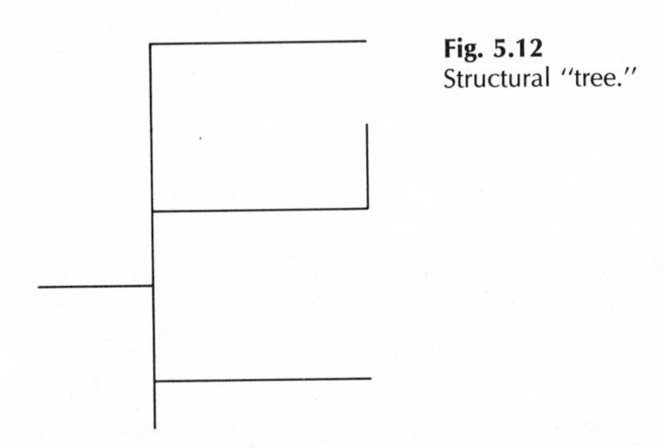

**Fig. 5.12**
Structural "tree."

Consider the structural part shown in Figure 5.12. It is called a *tree* because there are no closed loops included (trees in nature would look most uncommon if their branches grew back together at the tips!). The tree is shown with no internal hinges, and in this form it is easy to verify that if a set of external forces, in equilibrium themselves, are applied to the tree, the resulting internal forces can be determined by the equations of statics. We note that three equations of equilibrium can be written for the tree structure shown. If the tree were three-dimensional, there would be six such equations.

Now consider the rigid frame highway overpass shown in Figure 5.13

**Fig. 5.13**
Rigid frame highway bridge. *Bethlehem Steel*

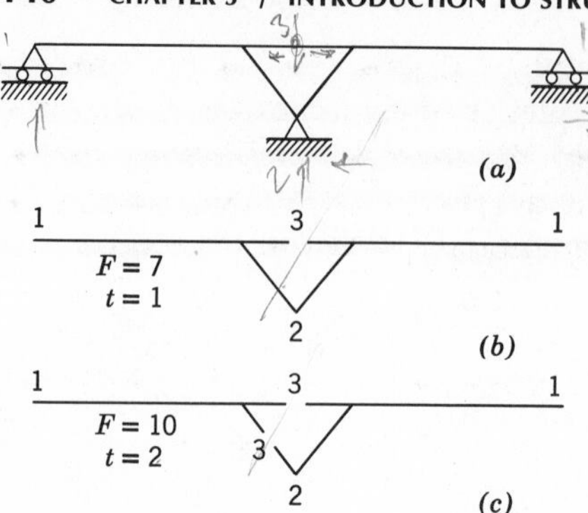


**Fig. 5.14**
Degree of static indeterminacy: rigid frame.

(idealized in Figure 5.14). The bridge has one closed loop and three support points. It is a relatively simple matter to cut the structure into a set of trees that have no internal hinges. Whenever a cut is made, we must introduce the appropriate number of unknown forces, noted on Figure 5.14*b* or 5.14*c*. Normally, there are three unknown forces in a solid bar in a planar frame: axial load, shear force, and bending moment. If the cut is at a hinge, where moment cannot be carried, only two unknown forces are introduced. A cut at each support introduces forces corresponding to each reaction constraint. Figures 5.14*b* and 5.14*c* indicate two of the many ways in which the structure can be cut into trees without hinges. The total number of unknown forces, $F$, and the number of trees, $t$, are shown on each figure. In Figure 5.14*b*, the structure has been cut into just a single tree. A total of seven forces were introduced. We have $3t = 3$ equations of equilibrium. The structure is therefore statically indeterminate to the 4th degree ($n = F\text{-}3t = 7\text{-}3$). In Figure 5.14*c*, the structure has been cut into two trees. There are 10 unknown forces and $3 \times 2 = 6$ equations of equilibrium, leaving us with the same conclusion as before. Obviously, the answer will be the same regardless of the way in which the structure is cut. The important parts of the above analysis are the creation of the trees, and an accurate count of the number of trees and the number of unknown forces.

Beam spans are particularly easy to study in this way. If the structure is cut at each internal hinge, and if each reaction constraint is severed, each segment will be a tree. Simple counting is all that is required for the evaluation. Figure 5.15 illustrates the determination of $n$ for a continuous beam.

Trusses may be studied in the same manner, but the approach is simplified if the tree is permitted to have hinges. All truss members are cut at midlength to form trees consisting of the cut members framing into a

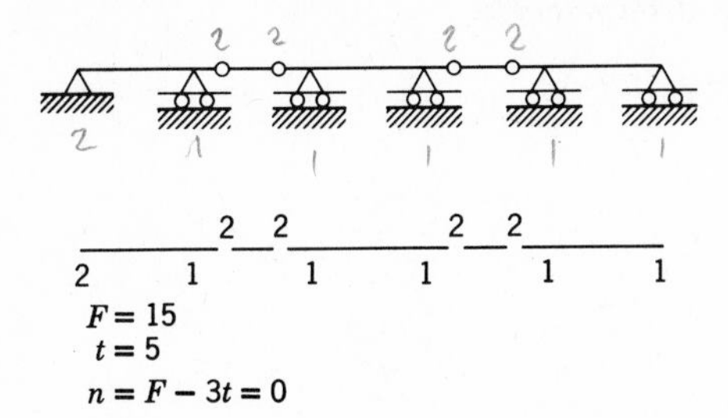


**Fig. 5.15**
Degree of static indeterminacy: beam.

single joint. Assume that a truss has $m$ members, $j$ joints, and $R$ reaction constraints. Then we have $j$ trees, each consisting of a set of members with concurrent forces meeting at the joints. Only two equations of equilibrium can be written for each tree because the moment equation is inapplicable where the forces are concurrent. There are $m+R$ unknown forces and $2j$ equations of equilibrium. The degree of static indeterminacy is therefore

$$n = m + R - 2j \tag{5.5}$$

This result is valid for all stable planar truss structures. Figure 5.16 indicates the computation of $n$ by Equation 5.5. The argument may be extended to the three-dimensional case to give

$$n = m + R - 3j \tag{5.6}$$

The degree of static indeterminacy $n$ can easily be computed, but as we mentioned before, the question of determinacy must await a study of geometric stability. Only a geometrically stable structure may be classed as statically determinate or indeterminate.

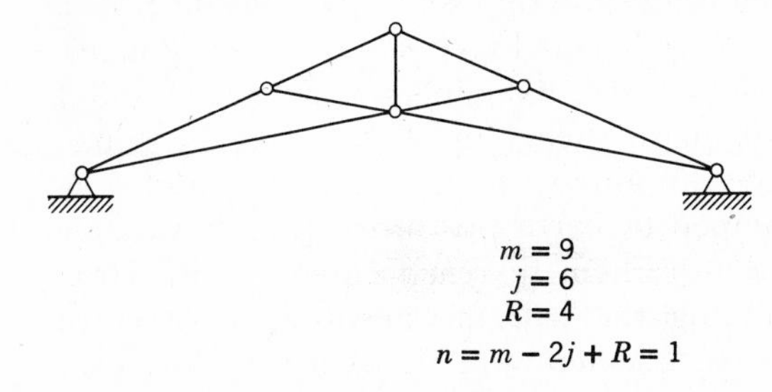


**Fig. 5.16**
Degree of static indeterminacy: truss.

## Geometric Stability

A K-truss system was recently proposed for a high-voltage transmission line tower. The lower part of one side of the tower looked something like the form shown in Figure 5.17$a$. Notice that the designers depended on support from the earth to resist horizontal loads resulting from applied

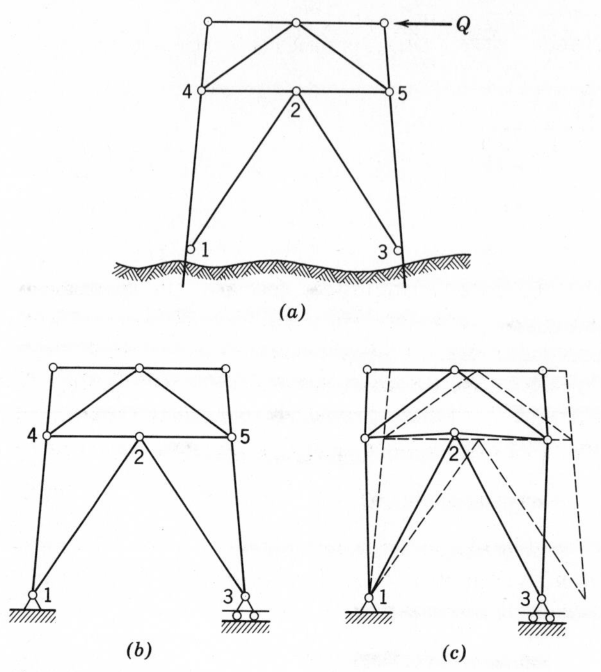


**Fig. 5.17**
K-truss as tower base.

wind forces, represented as $Q$ in the figure. If we examine the form shown in the figure, we may conclude that the bar forces in members 1-2 and 2-3 must be of equal magnitude and opposite sign. This is because joint 2 must be in equilibrium, and only equal forces in the bracing members can give zero resultant vertical force. Examination of joints 1 and 3 leads to the conclusion that the base shears must therefore be equal, neglecting the horizontal force components in members 1–4 and 3–5. Since the base shears are developed by earth pressure, a highly variable quantity, it may not be possible for them to reach equality until large deformations take place. Assuming the extreme in variation of earth pressure, the structure might be represented as in Figure 5.17*b*. This form is geometrically unstable, since it can change shape, as in Figure 5.17*c*, without stressing the members. When used properly, K-trusses are satisfactory but the structure of Figure 5.17 is useless. As designers, we must learn to avoid the choice of structural forms that are geometrically unstable. The addition of a single member from joint 1 to joint 3 removes the instability. Note that the idealization discussed above ignores bending effects in the legs.

Geometric instability is independent of loading and the strength and rigidity of members. We therefore are not concerned with particular loads, and for convenience we can imagine the members of the structure to be perfectly rigid.

We require a method for detecting an unstable structure early in the design stage. The basis for such a method lies in the fact that for a stable structure that is not statically indeterminate, the equations of equilibrium yield a unique solution. It follows that there must be at least the same number of unknown internal forces and external reaction components as there are equations of equilibrium. Therefore if $n$, the degree of static indeterminacy, turns out to be *negative*, we have fewer unknowns than equations, and the structure must be unstable. If $n$ is nonnegative, the structure may be stable or unstable. Our first task, then, is to consider the question of determinacy, noting the value of $n$, the degree of static indeterminacy. If this value is negative, the structure is unstable. No further analysis is needed in this case, for we have no interest in the analysis of unstable structures. The tower of Figure 5.17*b* falls in this category, with $n = -1$.

If $n$ is zero or positive, a careful study of the structure is required in order to detect possible unstable or critical forms. For beam and frame structures, a common sense approach without resort to quantitative analytical methods is useful. We simply examine the structure to determine whether any components may be unstable. We attempt to visualize possible mechanism forms of motion in either the whole structure or any of its parts. A *mechanism* is a structure or machine part with movable elements. In this regard it is often convenient to think of each tree configuration as the smallest subcomponent. The tree, with no internal hinges, is stable. If it is constrained to other trees in a stable manner, then stability is assured.

In order to consider the constraints on a tree component, it is useful to recall some concepts about forces.

1. A set of forces may be *parallel*, in which case there are no components in the perpendicular direction.
2. A set of nonparallel forces is *concurrent* if their lines of action pass through a single point.

If a structural component is constrained in a manner that allows only parallel or concurrent constraint forces, then that component is geometrically unstable. The reason for this is that nothing can restrain the component against motion in a direction perpendicular to the parallel constraints, and nothing can restrain the component against rotation about the point at which concurrent constraint forces meet. This means that for planar structures, at least three nonparallel, nonconcurrent forces must act to constrain the individual trees to other parts of the structure, since three independent equilibrium equations will be introduced when the tree is isolated.

The frame of Figure 5.18*a* is an example of a structure for which $n = 3$. Examination of the member *AC* indicates that a mechanism form is present. The free body diagram of *AC* (Figure 5.18*b*) indicates that regardless of applied load, the constraint forces *F* are parallel. Any other direction would impose an impossible condition: moment at hinge *A, B,* or *C*. Another way to consider the structure is to imagine a vertical load at *B*. Equilibrium of *B* requires member forces with a vertical component, and yet the members can only carry axial load because of the hinge connections at *A, B,* and *C*. Appreciable vertical displacements at *B* must take place in order to develop the necessary vertical components, and therefore the structure is unstable.

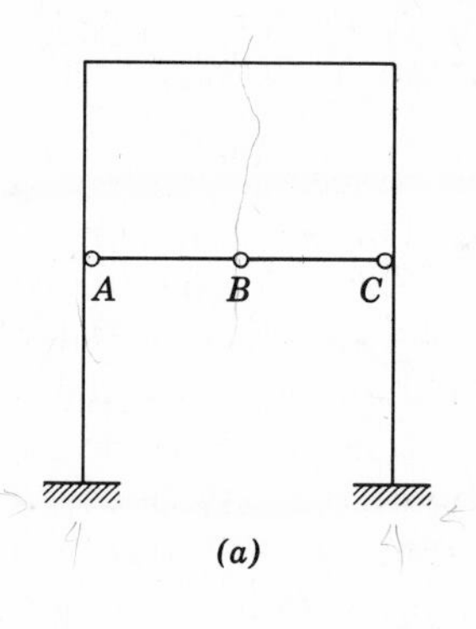


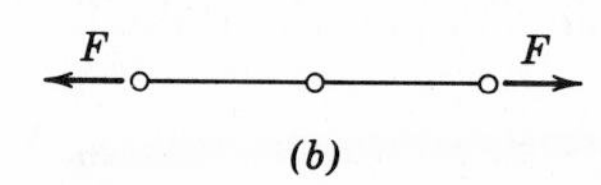


**Fig. 5.18**
Unstable frame.

In the case of trusses, the simplest approach is to consider the basic truss element, the triangle, and to study the ways in which triangles can be connected to form trusses. A single triangle is clearly rigid. A pair of members connected to two of the joints with their far ends connected to form another joint forms a stable system of two triangles. If the whole truss is built up in this way, it must be internally rigid. Such a structure is called a *simple truss*. When a simple truss is connected to its supports in a stable way, with a set of nonconcurrent, nonparallel reaction components, then the system is stable. Figure 5.19*a* illustrates the makeup of a simple truss. The photograph shows the construction of an aircraft assembly building for Lockheed Aircraft, which has clear area of 300 by 210 ft.

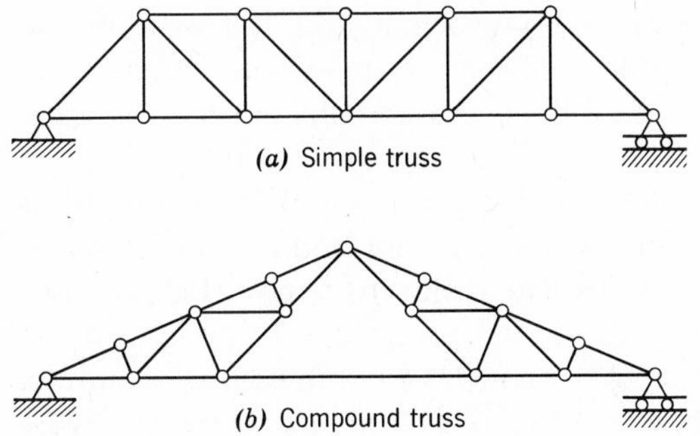

(*a*) Simple truss

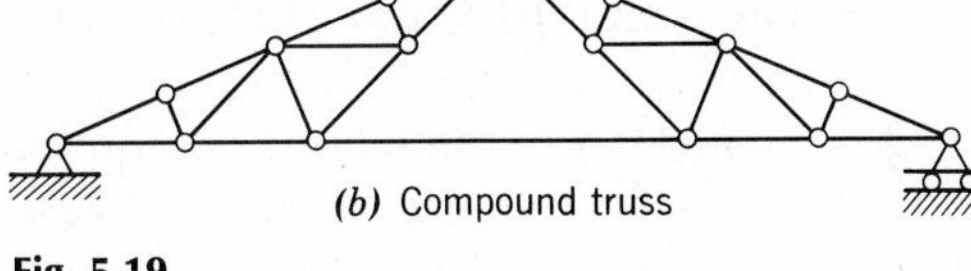

(*b*) Compound truss

**Fig. 5.19**
Truss forms.

*Bethlehem Steel*

If two simple trusses are connected with a set of bars or pin connections which provide nonconcurrent, nonparallel reactive components to each simple truss, then the system is stable. Such a system is termed a *compound truss.* Its identification is best performed by identifying the simple trusses as individual units, and then identifying the bars that provide the proper connections. The reaction components must of course be nonconcurrent and nonparallel. Figure 5.19*b* shows a compound truss.

Although the stability of simple and compound trusses can easily be verified by the examination of the triangles and their method of connection, a more general method is needed for *complex trusses.* A complex truss is one that cannot be classified as simple or compound. Complex trusses are not often used, and the following discussion would be useful only in rare cases except for the fact that it provides some insight into more general methods of determination of system stability. Complex trusses for which $n = 0$ may be analyzed for the presence of unstable or critical forms by the *zero load test.* This interesting method, suggested by Möbius, and explored in detail in Timoshenko and Young [1965], is based on some principles of linear algebra as applied to the physical performance of structural systems. A stable structure, for which $n = 0$, must have a unique set of member forces for any particular load condition. If no unique solution can be found (either no solution or more than one), the structure cannot be stable.

In terms of linear algebra, the determinant of the coefficient matrix of the equilibrium equations must be positive (the matrix must be positive definite) if a unique solution is to result. In order to examine stability, we could write the equilibrium equations and examine the determinant, but this would involve a great deal of labor unless performed by a computer. Incidentally, computer programs used in analysis often check for geometric stability in this way. The zero load test is based on the fact that more than one equilibrium solution for a very simple load case, namely zero load, is a definite indication of an unstable structure, when $n$ (degree of indeterminacy) $= 0$.

When zero loads are imposed on the structure, one obvious solution is zero member forces in all members. We need only find one other set of member forces that satisfy equilibrium in order to verify the presence of a critical (i.e., unstable) form.

The truss structure of Figure 5.20 provides an interesting example. It is a complex truss for which $n$ is zero. At first glance, then, the truss is statically determinate. You would find it interesting (but very frustrating) to attempt to write out the 12 equilibrium equations in terms of any arbitrarily chosen load system.

Application of the zero load test is simple in this case. For zero applied loads, the reactions at 5 and 6 are zero. Imagine a one kip tension in bar 1-4. Equilibrium of point 1 then requires a one kip compression in bars 1-6 and 1-2. Working clockwise around the truss, equilibrium of point 2 requires one kip tension in bar 2-5 and one kip compression in bar 2-3. Continuing to each joint, we find that everything is in equi-

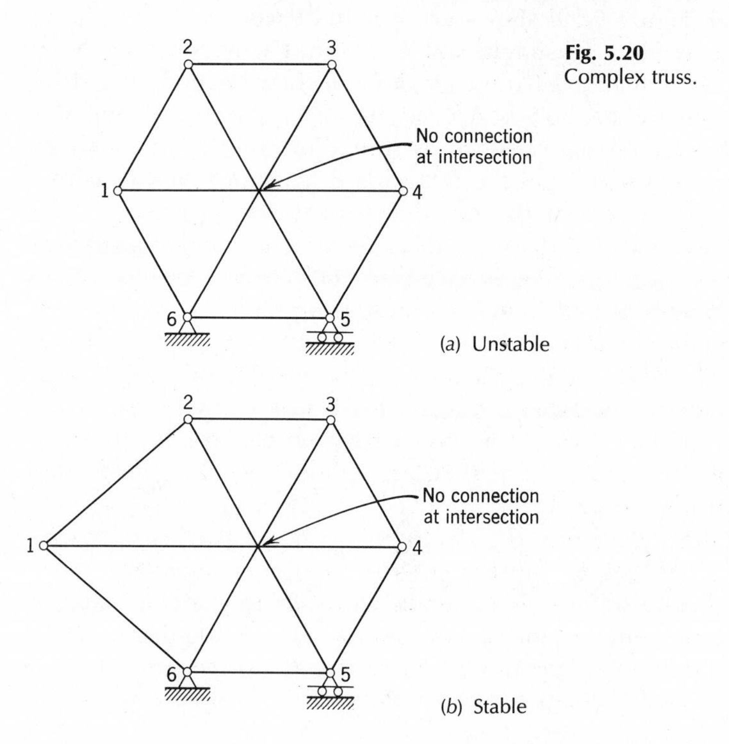


**Fig. 5.20**
Complex truss.

librium, with unit tension on the internal bars and unit compression on the external bars, and with zero applied loads. Yet with the same applied loads, the condition of zero force in all bars is also a legitimate equilibrium solution. Since $n = 0$ and we have more than one solution, the truss must be unstable. A model will readily verify this conclusion. You may wish to compare the analysis of the truss of Figure 5.20*b* with that of 5.20*a*.

The two items of suggested reading should be consulted by those wishing to pursue topics of determinacy and stability. Timoshenko and Young are particularly strong on trusses including the zero load test, while Norris and Wilbur cover the subject from a more general viewpoint.

## 5.4 SUMMARY

The purpose of this chapter has been to introduce a number of important concepts of structural analysis before actual considerations of analysis are met in detail. You will find that many of these ideas will become much clearer as a better understanding of structural analysis develops with experience and practice. Questions of linearity and elastic versus inelastic behavior, the use of superposition, and of determinacy and geometric stability, will occur often in future study of structural behavior. These questions must be carefully asked and answered whenever the analysis of any new structural type is begun.

### *Suggested Reading*

Timoshenko, Stephen P., and Young, Donovan H. [1965]: *Theory of Structures,* 2nd ed., McGraw-Hill, New York, pp. 85-98.

Norris, C. H., and Wilbur, J. B. [1960]: *Elementary Structural Analysis,* 2nd ed., McGraw-Hill, New York, pp. 60-85.

## PROBLEMS

5.1 (*a*) Refer to Figure 5.5*a* and Equation 5.1. Compare the displacement $D$ due to a 3-k force with that due to a 6-k force. Let $A = 2$ in.$^2$, $s = 6$ in., $l = 120$ in., and $E = 30{,}000$ ksi.

(*b*) Refer to Figure 5.5*d* and Equation 5.2. Use the values given in (*a*), except let $s = 48$ in.

(*c*) At what point in the development of Equations 5.1 and 5.2 does the assumption of linear *materials* arise?

(*d*) Compare the two equations, and discuss the situation when $s$ is intermediate between the cases given.

5.2 Compare the behavior of an imperfect column under axial compression (Figure 5.3) and the two-bar truss (Figure 5.6). Which type of behavior seems most desirable?

5.3 Determine the reactions $H_A$, $V_A$, $H_G$, $V_G$, for the frame of Figure 5.9. $P_1 = P_3 = P_5 = 0$, $P_2 = 5$ k, $P_4 = 10$ k. Use the condition of symmetry to simplify the analysis. Assume a span of 40 ft and a rise of 30 ft. Loads $P_2$ and $P_4$ are 10 ft from $D$.
*Hint.* Make up a symmetric load condition and an antisymmetric

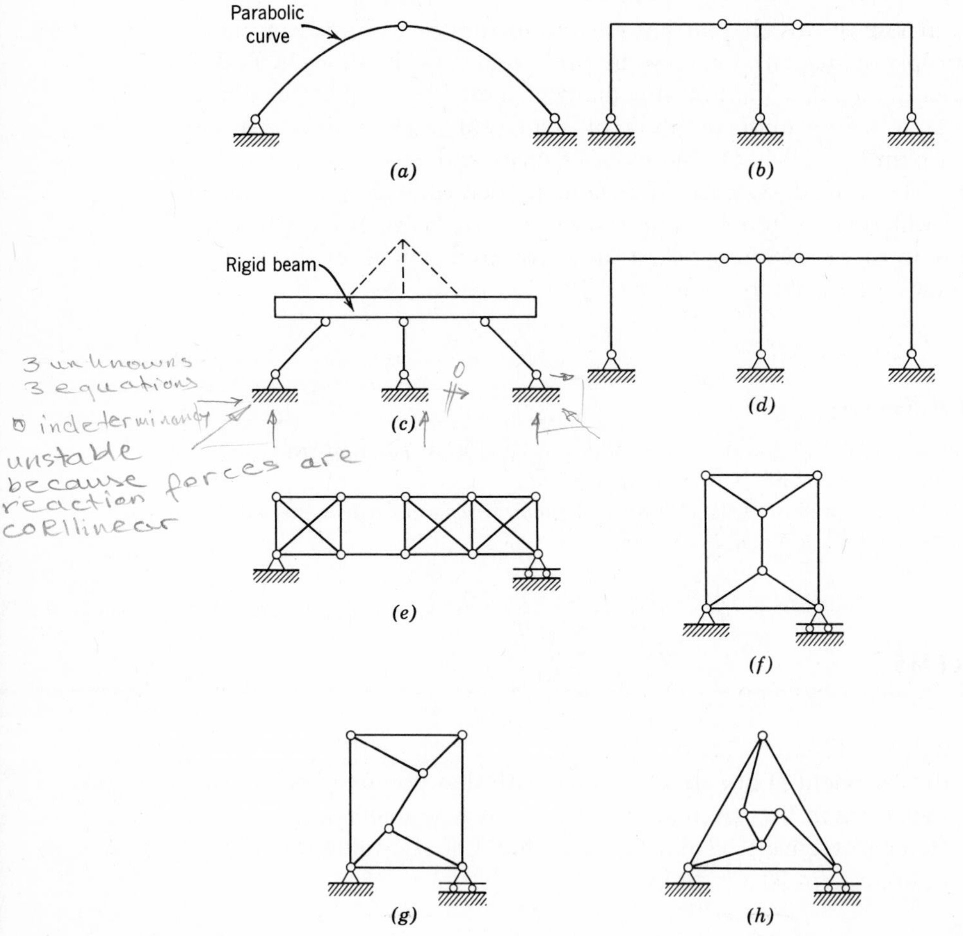


**Fig. P5.4**

load condition which, when added together, yield the actual loading. Draw free body diagrams of each half of the structure. For a symmetric structure symmetrically loaded, the center pin cannot carry a vertical force, since any such force would destroy the symmetry. For a symmetric structure loaded antisymmetrically, the center pin cannot carry a horizontal force since this would destroy the antisymmetry.

5.4 Classify the structures (*a*) to (*h*) in Figure P5.4 as to determinacy and stability. Justify your answers. In the case of trusses, also note the type of truss.

5.5 Can a nonlinear structure be statically determinate?

5.6 Discuss the behavior of the column of Figure 5.2, assuming that it was loaded in tension instead of compression. Can you comment on the relative safety of such forms?

5.7 The Solleks Bridge is indicated in Figure 1.1*a*. Discuss geometric stability and static determinacy for (*a*) The condition when the hinged connections exist in the girder (before deck slab poured), and (*b*) the finished condition, in which the girder hinges are eliminated (struts are still pin-connected to girders).

5.8 Consider the three-dimensional structure of Figure P5.8. Verify that it is stable, and find its degree of static indeterminacy. *Hint.* Keep in mind the six basic internal force quantities that are possible, and develop an extension to the discussion regarding trees and the number of equations and unknowns concerned.

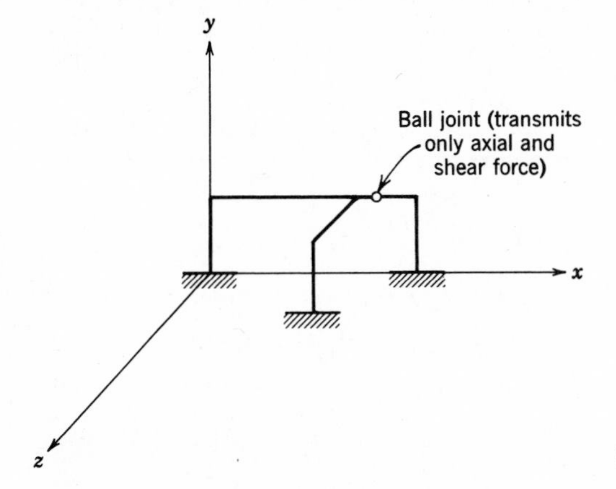


**Fig. P5.8**

5.9 Comment on geometric stability of the forms shown in Figures 4.4 and 4.10. Under what conditions are geometrically unstable structures useful?

5.10 (*a*) Determine the degree of static indeterminacy for the frame shown in Figure 7.13 (*b*) Assume that X-bracing is installed in the five right side panels. The X-bracing is pin connected to the corners of the panel. What is the degree of static indeterminacy? *Hint:* Admit trees with internal hinges; each cut bar with hinged ends has only a single unknown axial force.

CHAPTER 6

# Analysis of Statically Determinate Structures

Pavilion, Raleigh, N.C. *Bethlehem Steel*

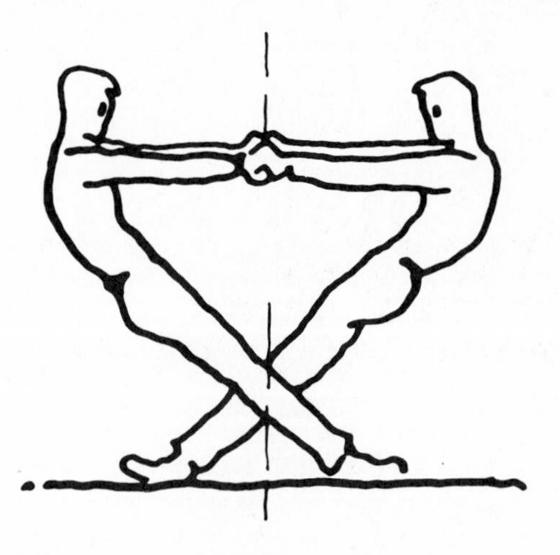

# 6

## Analysis of Statically Determinate Structures

The calculation of the actions or forces (axial loads, shears, bending moments, and torsional moments) in every member of a structure is an important part of the design process. As we have seen in Chapters 1 and 5, analysis should not be divorced from other phases of structural design.

In Chapter 5 we learned how to identify statically determinate structures; here, we shall study their analysis. We shall emphasize the underlying principles involved and the action or behavior of structures rather than the slight differences in arithmetical approach for the various types of structures. The subject matter of this chapter is of fundamental importance to the structural engineer. It is precisely his understanding of the play of forces in structures that enables the structural engineer to take a key role in all phases of an engineering project.

In statically determinate structures, the unknown internal forces and reactions can be calculated by considering the equilibrium of the entire structure or of portions of the structure. As opposed to this, indeterminate structures contain excess reaction components or members (see Section 5.3); therefore, the number of unknowns exceeds the available number of equilibrium equations. We shall analyze indeterminate structures in Chapter 7 and in Volume 2 by considering deformations in addition to equilibrium.

The fact that equilibrium conditions are of paramount importance in structural analysis is illustrated by the fact that the word "statics" or its derivatives connotes the entire field of structural theory as well as the analyst (e.g., "statik" and "statiker" in German, Turkish, and some other languages, "statika" and "statikus" in Hungarian.)

### 6.1 FREE BODY DIAGRAMS

The analysis of all structures is based on the fact that the structure is in equilibrium under the action of the loads. The magnitudes of the reactions are such that the applied loads are exactly counteracted or resisted in accordance with Newton's third law. Furthermore, any part of a structure is also in equilibrium. This fact is used to determine internal forces in a structure.

***Example 6.1*** Let us remove and study one member from the Solleks River Bridge, shown in Figure 6.1*a*. The forces acting on this member are the gravity effects on the mass of the member (weight), the live load, the force $P_4$ transmitted by the center portion of the bridge, and the reactions at points 5 and 6 as shown in Figure 6.1*b*. Moments cannot act on the ends of the member because of the hinged support connections. Such a figure, which shows a portion of a structure under *all forces* acting on it, is called a *free body diagram*. The equilibrium analysis of a free body yields equations for the evaluation of the forces that are unknown.

The sketching of free body diagrams is of extreme importance in structural analysis. We can study the behavior of a structural element or subassembly by isolating it from the rest of the structure. In fact, as we shall see in Volume 2, modern analysis methods consider every member or even parts of members as *elements* and build up the entire structure by assembling these elements. Thus, many structural engineers have become accustomed to view every structure as an assemblage of free bodies connected by internal forces.

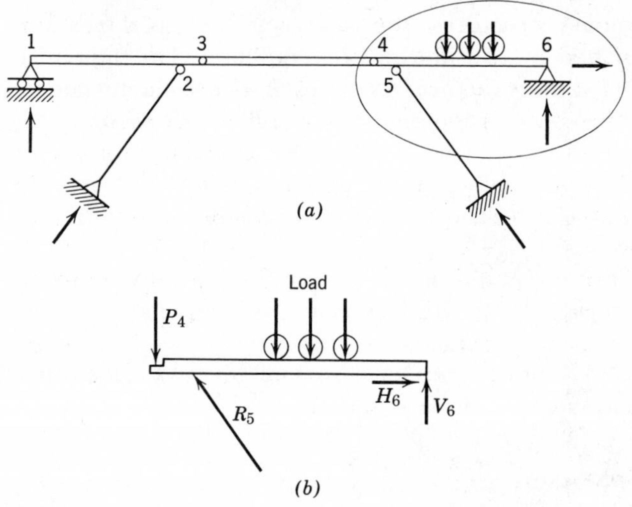


**Fig. 6.1**
Solleks River Bridge; illustration of free body diagram.

We find that students can eliminate many errors and avoid doing unnecessary work by making liberal usage of free body diagrams. In many instances, free body analysis serves as a direct method for evaluating or checking internal forces or reactions. In other cases, the study of the equilibrium or the deflection of a free body element reveals much about the action of various types of structures.

***Example 6.2*** Study the structural action of a nutcracker (Figure 6.2). This simple example illustrates that the separation of the three elements of the "structure" directly yields the forces at the joint; that is, the forces exerted by the pin on the members and the forces acting on the pin.

Instead of considering more examples of free body diagrams here, we shall use this obvious but very important technique throughout this text. We urge you to acquire the habit of separating structures into free body diagrams as a means of analysis and as an aid in visualizing the transmission of forces in structures.

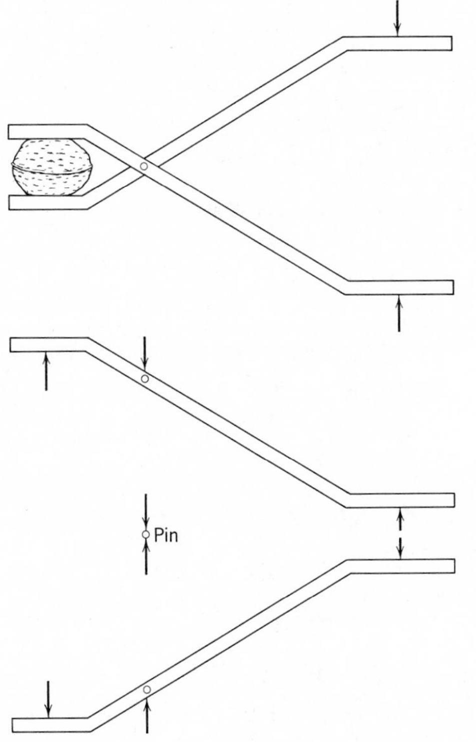


**Fig. 6.2**
Free body diagrams for a nutcracker.

## 6.2 EQUILIBRIUM ANALYSIS

Consider any stationary structure or object acted on by a series of forces, including external loads, reactions, and body forces caused by the weight of the elements. The conditions of equilibrium are best established with reference to a fixed coordinate system, such as orthogonal $x, y, z$ axes (Figure 6.3*a*). It is usually convenient to replace all forces by their components along the chosen reference axes.

The condition of equilibrium in the $x$ direction expresses the fact that there is no net (or unbalanced) force acting in that direction which would accelerate the structure or element. Thus the algebraic sum of all forces along the $x$ axis must be zero, or we can express it mathematically as $\Sigma F_x = 0$. Similar equations hold along the $y$ and $z$ axes. Three additional equations of equilibrium state the fact that the structure or element does not spin about any of the three axes due to unbalanced moments. The satisfaction of the three force equations and the three moment equations guarantees that the structure, as an arbitrary body in space, is in static equilibrium; in fact, this is a necessary and sufficient condition.

The six equations of equilibrium are:

$$\begin{aligned} \Sigma F_x &= 0 \qquad \Sigma M_x = 0 \\ \Sigma F_y &= 0 \qquad \Sigma M_y = 0 \\ \Sigma F_z &= 0 \qquad \Sigma M_z = 0 \end{aligned} \tag{6.1}$$

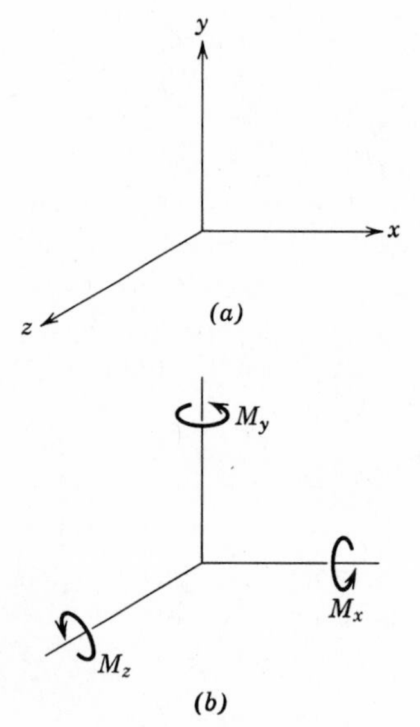


**Fig. 6.3**
References axes for equilibrium equations.

The notation $\Sigma M_x$ signifies the moment of all forces and concentrated moments about the $x$ axis, as indicated in Figure 6.3*b*.

We can generalize these static equilibrium conditions to include dynamic effects. Whenever a structure moves or rotates at nonuniform speed, inertia forces are created as a result of the linear or angular acceleration. These forces are of the form $m\ddot{x}$ or $J\ddot{\Theta}$ where $m$ is the mass of an element, $\ddot{x}$ is its acceleration, $J$ is the mass moment of inertia, and $\ddot{\Theta}$ is the angular acceleration. The displacements $x$ and $\ddot{\Theta}$ are functions of time and the corresponding forces throughout the structure also vary with time. Such dynamic forces are important in the design of aircraft, missiles, and machine parts, and of most structures subjected to earthquake and wind loadings.

The primary use of equilibrium analysis is to evaluate the reactions and the internal forces by forming a series of free body diagrams. If a force acts in an arbitrary direction with respect to the coordinate axes, we replace the force with its components along the three coordinate axes. In the case of a general three-dimensional structure, we need all six equilibrium equations. However, many three-dimensional structures can be idealized as a series of two-dimensional components with loading only in the plane, such as the plane frames of a building.

A planar structure is a structure or member in the $x$, $y$ plane with no force acting in the $z$ direction nor moments (torques) about the $x$ and $y$ axes. The $M_z$ moment then represents moments about the $z$ axis or any point in the plane. Thus, for planar structures, we have only three equations of equilibrium; we shall give an example of the use of these equations at the end of this section.

If all forces acting on a two-dimensional structure are parallel, for example in the same direction as the $y$ axis, then the condition $\Sigma F_x = 0$ contains no terms. There are then only two effective equations ($\Sigma F_y = 0$ and $\Sigma M_z = 0$) for this common type of loading.

Several other special cases can occur. For example, if all forces located in a plane pass through a point, the summation of moments about this point would not contain any terms and only two equations of equilibrium are necessary.

Instead of following a rigid sign convention for forces and moments, we shall adopt various coordinate directions and positive directions of forces to suit the problem at hand. For trusses, it is customary to assume tensile force as positive. The sense of the unknown force components are assumed and if the calculation happens to yield a negative number, it simply means that the actual direction of the force is opposite to the one assumed.

***Example 6.3*** Separate the main members of the basic frame of the Solleks River Bridge, which was described in Chapter 1, as free bodies and determine the forces acting on them as a result of the construction load shown on a line drawing in Figure 6.4*a*. The 2.5% slope of the bridge will be neglected here in calculating the forces. The loaded member 3-4 is supported at the internal hinges.

The free body diagram of member 3-4 is shown in Figure 6.4*b*. The end shears (which can be thought of as reactions to this member) at 3 and 4 are calculated by taking moments about 3 and 4. We get

$$\Sigma M_4 = 75V_{3,4} - 55\ (22) = 0$$

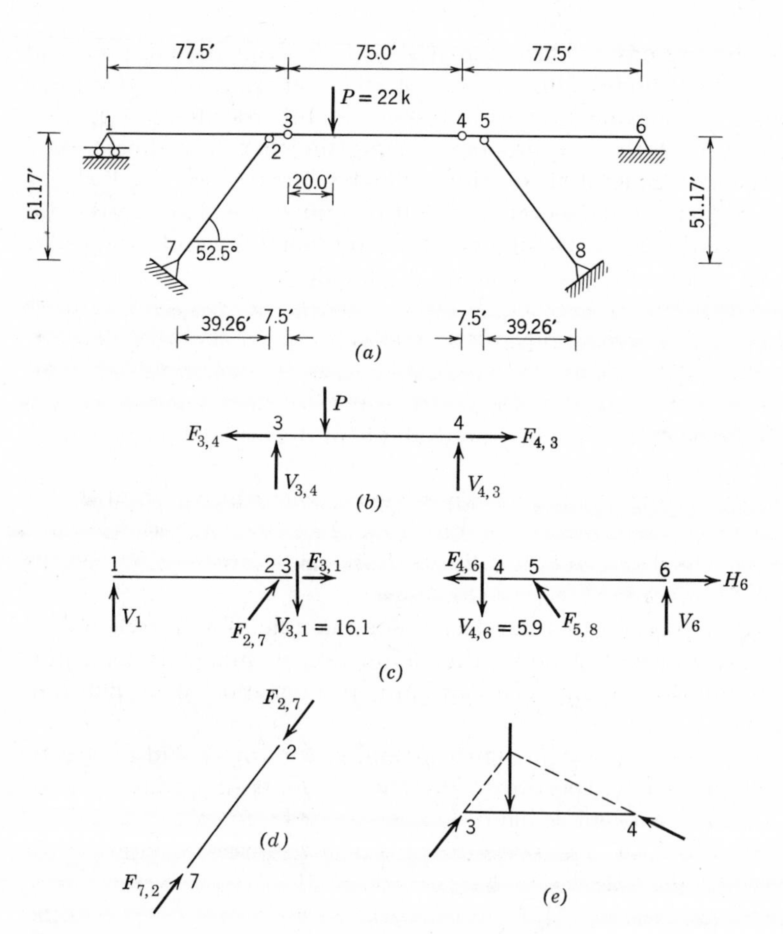


**Fig. 6.4**
Equilibrium analysis of Solleks Bridge.

from which

$$V_{3,4} = 16.1 \text{ k}$$

and

$$\Sigma M_3 = 20\ (22) - 75V_{4,3} = 0$$

from which

$$V_{4,3} = 5.9 \text{ k}$$

where the first subscript denotes the joint number where the force acts and the second subscript specifies the other end of the member. Obviously, $V_{3,4} + V_{4,3} = P = 22$ because of vertical equilibrium. We cannot yet evaluate the axial forces $F_{3,4}$ and $F_{4,3}$ but we know that they must balance each other.

The end forces on members 3-4 are transmitted through the pins at 3 and 4 to the members 1-3 and 4-6, as shown in Figure 6.4*c*. Examination of the forces acting on the pin at 3 reveals that if a force acts to hold up member 3-4 at 3, a force of the same magnitude must act down on member 1-3 at 3. In other words, the pin at 3 is in equilibrium under the opposing forces $V_{3,4}$ and $V_{3,1}$ vertically and $F_{3,4}$ and $F_{3,1}$ horizontally.

Member 1-3 is supported at 1 and at 2. Isolation of the "pin-ended" member 7-2 (Figure 6.4*d*) shows that the loads at its two ends must be collinear (act along the same line). This general observation follows readily from the fact that, in the absence of other forces on the member, neither of the end forces can have projections normal to the member because that condition would produce moment about the ends of the member. Thus the reaction of member 1-3 at 2 must be along the direction of 7-2, as indicated in Figure 6.4*c*.

The roller at 1 can support only a vertical force $V_1$. The value of $V_1$ is best determined by taking moments about point 2; since the lines of action of the other unknowns intersect at this point, these forces are eliminated from the equation. Referring to the left free body of Figure 6.4*c*

$$\Sigma M_2 = 70V_1 + 7.5\ (16.1) = 0$$

from which $V_1 = -\ 1.7$ k. (The negative sign indicates that this force acts opposite to the assumed direction, i.e., downward.) Vertical equilibrium of member 1-3 demands that the vertical component of the force at 2 must be $V_2 = 16.1 + 1.7 = 17.8$ k. Note that we could also have obtained this force by taking moments about 1 or 3. Therefore, the force in member 7-2 is

$$F_{2,7} = F_{7,2} = V_2/\sin 52.5 = 17.8/0.793 = 22.4 \text{ k (compression)}$$

The horizontal component of this force is

$$X_2 = 22.4(\cos 52.5) = 13.6 \text{ k}$$

which exerts a longitudinal force on member 1-3 at point 2, acting to the right. This indicates that the segment 2-3 of member 1-3 is under compression. Horizontal equilibrium at 3 dictates that the axial force in the main girders is

$$F_{3,1} = F_{3,4} = F_{4,3} = -13.6 \text{ k (compression)}$$

Now we examine the right half of the bridge (Figure 6.4*c*). The pin at 4 receives the downward force $V_{4,6} = V_{4,3} = 5.9$ k and the axial compressive force $F_{4,6} = F_{4,3} = -13.6$ k. The hinged support at 6 can take both horizontal and vertical forces. We can evaluate the latter by taking moments about 5:

$$\Sigma M_5 = -7.5\ (5.9) - 70V_6 = 0$$

from which $V_6 = -0.6$ k (downward).

Note that the reactions at 1 and 6 are downward and vertical anchors would be required. Actually, the self-weight of the "anchor-spans" (1-2 and 5-6) was sufficient to counteract the uplift due to the actual loads on the Solleks River Bridge.

Summation of moments about 6 gives the vertical component in member 5-8 as $V_5 = 77.5 \times 5.9/70.0 = 6.5$ k. The force in member 5-8 is $F_{5,8} = 6.5/\sin 52.5 = 8.2$ k, which has a horizontal component $H_5 = 8.2(\cos 52.5) = 5.0$ k, acting to the left at joint 5. The axial force in member 4-6 is 13.6 k compression between 4 and 5 and $13.6 - 5.0 = 8.6$ k compression between 5 and 6. Thus the horizontal component of the reaction at 6 is $H_6 = -8.6$ k to the left and a restraint would be required.

The reexamination of the equilibrium of member 3-4 (Figure 6.4*e*) shows an interesting situation. If the horizontal and vertical components at each of joints 3 and 4 are combined, only three forces act on this member. These three forces intersect at a common point. This is the case whenever only three forces act on a planar structure or member since otherwise one force would have an unbalanced moment about the intersection of the other two forces.

The entire bridge is redrawn in Figure 6.5 with all the forces shown. Where appropriate, pairs of orthogonal forces are shown with their resultants as dashed lines. As a check of the arithmetic, let us examine the overall equilibrium of the structure.

$$\Sigma F_x = 13.6 - 5.0 - 8.6 = 0 \quad ✓$$

$$\Sigma F_y = -1.7 + 17.8 - 22.0 + 6.5 - 0.6 = 0 \quad ✓$$

$$\Sigma M_1 = +\,97.5\ (22) + 230\ (0.6) - 22.3\ (70)\ (\sin 52.5) - 199.3\ (6.5) + 51.17\ (5.0) = 0. \quad ✓$$

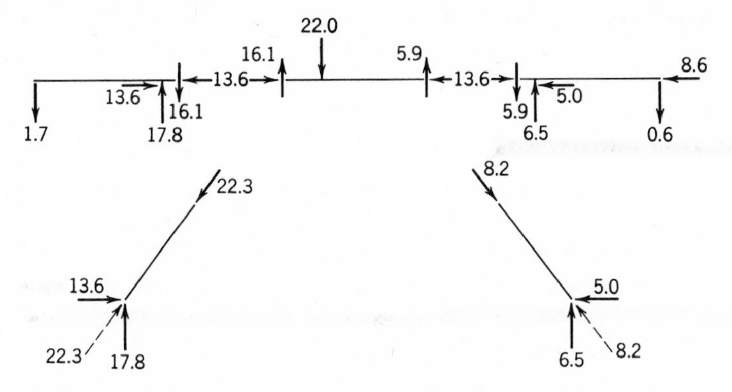


**Fig. 6.5**
Results of Example 6.3.

The center span (3-4) of the Solleks River Bridge is suspended at the ends of the anchor spans, which cantilever inward. This method of construction of long-span bridges is quite common. The basic structural action of a large cantilever bridge truss is illustrated in Figure 6.6.

In summary, the equilibrium analysis of statically determinate structures consists of the following steps: (1) free body diagrams are drawn of the components that make up the structure; (2) forces and moments are defined where they can exist at all cuts; (3) the equilibrium equations are applied to all free bodies, including the entire structure, in such an order that the number of unknowns is minimized in the equations.

**Fig. 6.6**
Illustration of cantilevered construction, the Forth Bridge. *W. Westhofen*

We shall examine the reactive and internal forces in several types of structures in the following sections of this chapter. We shall follow the same order as in the discussion of structural forms in Chapter 4.

## 6.3 AXIAL FORCE STRUCTURES

### Cables

Cables are frequently used to support loads over long spans, such as in the case of suspension bridges and roofs of large, open buildings. The only force in a cable is direct tension since cables are too flexible to carry moments. Consider the suspension cable shown in Figure 6.7*a*; the concentrated loads are caused by vertical hangers on which the bridge deck is supported. The forces in the cable segments depend on their directions: at points 2, 3, and 4, the applied loads are balanced by the cable forces as shown in Figure 6.7*b*. The shape of the cable with a certain length is therefore uniquely determined from equilibrium conditions. Knowing one coordinate, such as the sag at 2, we can calculate the sag at any point.

There are four unknown reaction components. This is a planar structure and therefore there are three equations of equilibrium for the whole

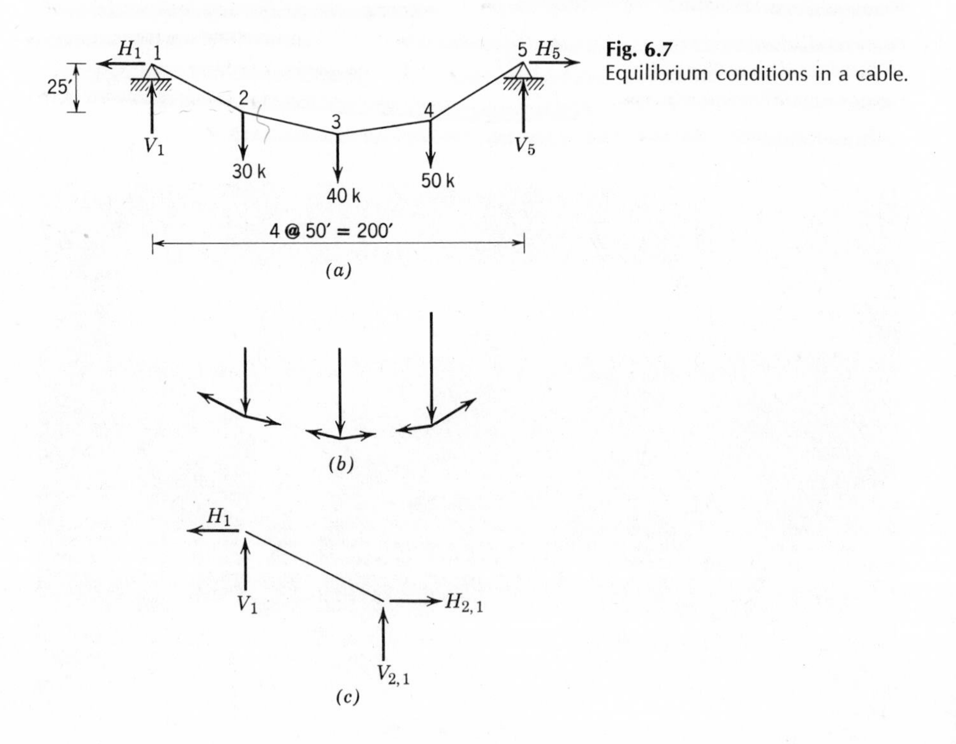


**Fig. 6.7**
Equilibrium conditions in a cable.

structure as a free body. A fourth equation follows from the fact that the moment is zero at any point on the cable. The first equilibrium equation may be taken as

$$\Sigma M_5 = 200V_1 - 150\ (30) - 100\ (40) - 50\ (50) = 0$$

from which

$$V_1 = 55.0\ \mathrm{k}$$

Note that we have arbitrarily, but with some foresight, assumed the directions of the unknown forces instead of taking them positive along the direction of some specified coordinate axes. The advantage of this method is that we can attempt to visualize or guess the direction of the unknowns, and the sign of the results tell whether the assumptions were correct. In this manner, one can develop a feeling for the direction of the forces and sometimes detect numerical errors. We should take moments about a point along the line of one of the unknowns or, even better, at the intersection of two unknowns, in order to eliminate unknowns from the equations and thus avoid simultaneous equations.

Since the cable cannot carry moments, we can cut through it at any point, such as just to the left of 2, creating a free body diagram of part of the cable (Figure 6.7*c*). We then sum moments about the cut end, point 2 in the present case

$$50V_1 - 25H_1 = 0$$

from which

$$H_1 = 50 \times 55/25 = 110\ \mathrm{k}$$

Examination of this free body reveals that the resultant of the two reaction components at 1 must have the same direction as the cable segment 1-2. In general, the resultant end forces on an unloaded member with hinged ends must be collinear since otherwise an unbalanced moment would exist, which would be a contradiction (compare with members 2-7 and 5-8 for the Solleks River Bridge, Figure 6.4).

We can use the second equilibrium equation in order to determine the vertical reaction component $V_5$ at the right reaction.

$$\Sigma F_y = V_1 - 30 - 40 - 50 + V_5 = 0$$

hence, $V_5 = 65$ k.

Instead of $\Sigma F_y = 0$, we could have summed the moments about a point, for example $\Sigma M_1 = 0$ to obtain $V_5$. Two moment equations can always be substituted for one moment and one force equilibrium equation, or the alternative equation may serve as a check of the arithmetic.

The third, yet unused, equation of equilibrium

$$\Sigma F_x = 0 \text{ gives } H_5 = H_1 = 110 \text{ k}.$$

Having determined the four reaction components, we can now evaluate the internal forces. The shape of the cables depends on the forces (this is a form-active structure); thus the sags of the load points must also be calculated. Considering the free body to the right of point 4 and taking moments about 4 gives $\Sigma M_4 = 110y_4 - 50V_5 = 0$; hence the sag at point 4 is $y_4 = 29.5$ ft. Similarly, summation of moments about 3 of the forces acting on the free body *either* to the left or to the right of point 3 gives the sag at the center: $y_3 = 36.4$ ft. The equilibrium conditions at points 2, 3, and 4 dictate the magnitude of the internal cable forces acting in the known directions. Since the forces on the free bodies at each of those points go through a point (Figure 6.7*b*), we have only two equations ($\Sigma F_x = 0$ and $\Sigma F_y = 0$) to evaluate the cable forces. The results are shown in Figure 6.9.

**Fig. 6.8**
Girder of Solleks River Bridge transported by cable.
*ABAM Engineers, Inc.*

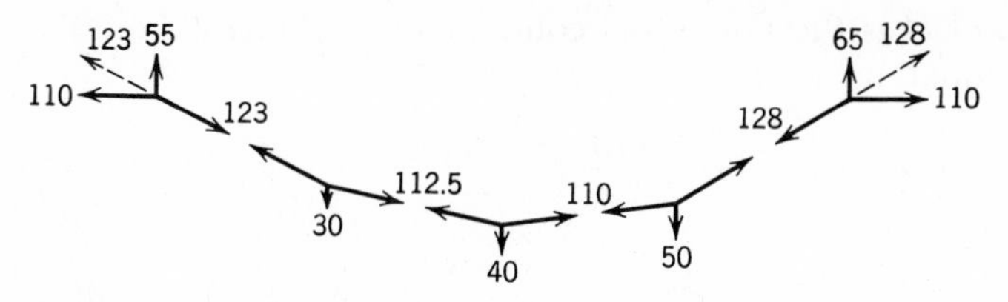


**Fig. 6.9**
Forces in the cable structure of Fig. 6.7.

It is easy to see how the loads and the shape of the cable interact to produce a structure that is in equilibrium under tensile forces only. The angle change at 4 is greater than at 2 in order to equilibrate the larger load there. Observe that we did not have to solve simultaneous equations to evaluate the four reaction components because we could use the various equilibrium equations in such an order that each contained only one unknown. Sometimes this is not possible; for example, if the two supports (points 1 and 5) were at different elevations, the horizontal reaction components would enter as second unknowns in the moment equations and then at least two equilibrium equations would be coupled (see Example 6.4 and Problem 6.10).

There is an abrupt change in direction in the cable at every load point so that the cable forces can balance the load; obviously in practice the load is not applied at an ideal point but over a distance (e.g., through a hanger), and correspondingly instead of a sharp break we have a gradual curving of the cable under the load. The number of breaks in the cable geometry increases with the number of loads. At the limit, the cable is a continuous curve under a distributed load.

***Example 6.4*** Analyze the cable under uniform load shown in Figure 6.10.

The loading determines the *shape* of the cable but we need to know the sag at one point to be able to calculate the sag and forces at any other point; usually the sag $h$ at the center is given. We calculate the two components of the left reaction by equating the moments about points 2 and 3 to zero. For the left portion of the cable

$$\Sigma M_2 = V_1 L/2 - H(h - d/2) - (qL/2)L/4 = 0$$

and for the entire cable

$$\Sigma M_3 = Hd + V_1 L - (qL)L/2 = 0$$

where $h$ is the sag of the center point, measured from the chord connecting the end points, and $H$ is the horizontal component of the reaction. The solution of these equations is

$$H = \frac{qL^2}{8h} \tag{6.2}$$

and

$$V_1 = \frac{qL}{2}(1 - d/4h)$$

The free body equilibrium analysis of a portion of the cable shows that the horizontal component of the cable force is $H$ everywhere for gravity loads and is independent of the end elevations for constant $h$. Thus the right reaction also has $H$ as the horizontal component. The equilibrium condition in the vertical direction yields the fourth reaction component

$$V_3 = qL - V_1 = \frac{qL}{2}(1 + d/4h)$$

When the two supports are at the same elevation, $d=0$, and $V_1=V_3=qL/2$.

In order to find the shape of the cable, we write the equation of moments about a general point $P$ with coordinates $x$, $y$ (Figure 6.10)

$$V_1 x - (qx)\frac{x}{2} - H\left(y - \frac{x}{L}d\right) = 0$$

from which

$$y = \frac{4hx}{L}\left(1 - \frac{x}{L}\right) \tag{6.3}$$

This is a second-order parabola. The shape of a cable under nonuniform loading is not a second-order parabola. This dependence of the shape of a structure on the loading is a special feature of cables. The deflection of the cable under its own weight (as opposed to the uniform load $q$) is not ex-

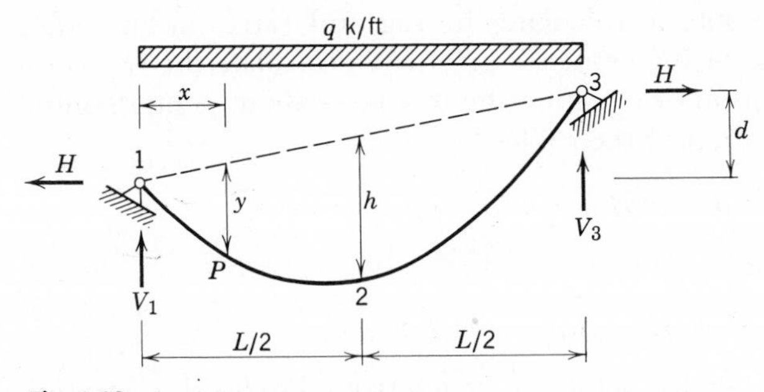


**Fig. 6.10**
Cable with uniformly distributed loading.

actly a parabola but rather a catenary or cosh function; however, in practice, the self-weight of the cable is much smaller than the superposed load and it is usually lumped with the latter. A transmission line assumes the catenary shape.

The maximum cable force occurs at the right support since the horizontal component $H$ is constant and the maximum vertical cable force component occurs at 3; thus $T_3 = (H^2 + V_3^2)^{1/2}$ is the design cable force for a cable with uniform cross section.

From Equation 6.2 we see that the sag of the cable is inversely proportional to the horizontal reaction components. Thus, flat cable structures exert very large forces on the supports (which can easily be demonstrated by putting weights on ropes with various sags). When the cable is unusually flat, the additional sag resulting from the elongation of the cable must also be considered (see Volume 2, Chapter 17 or Timoshenko and Young [1965]). Furthermore, such flat and flexible cable structures tend to vibrate under wind load and must be stiffened.

### Trusses

Trusses are another major type of structure that have only uniaxial (tensile or compressive) forces. We assume that the members of a truss are pin-jointed at their ends. This is not exactly the case because the members are, in most applications, joined with connections that take some moment. We also assume that the loads are applied only at the joints. The loads from a bridge deck are usually carried through stringer beams and floor beams to the joints (Figure 6.11). Thus, all truss members receive forces only through the joints at their ends and, therefore, these two end forces must be collinear and opposite to be in equilibrium. Finally, we assume that the centroidal axes of the various members framing into a joint will intersect at a common point.

There are two common ways of calculating the forces in the members of a truss. We can bisect the structure and write equilibrium equations for one side of the cut truss (as a free body). The forces in the members that are cut can then be calculated if three or fewer members with unknown forces are cut: this is called the *method of sections.* In the other approach, we isolate each of the joints in turn and examine its equilibrium under the action of the load and the known and unknown member forces. If the joints are considered in the proper order, there will usually be only one or two unknown member forces at each joint to be calculated. This approach is called the *method of joints.*

The method of joints is effective if we want to calculate all forces in a

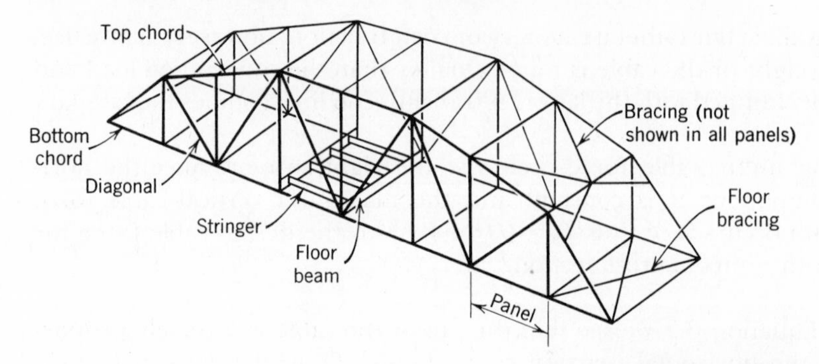


Fig. 6.11
Typical members in a truss bridge.

truss, but the method of sections is obviously superior if we seek the forces only in some of the members. In such a case, sections can be made only through the selected members, whereas the method of joints would require the analysis of joints from one end of the structure progressively up to the particular members.

***Example 6.5*** Calculate the forces in members 5-7, 5-6, and 4-6 of the 128-ft long bridge truss shown in Figure 6.12*a*. The width of the bridge is 36 ft. This is called a Warren truss with sloping top chord. The joint loads result from a total deck weight of 160 psf, which includes the weight of the concrete slab, surfacing of the roadway, sidewalks, structural elements, and utilities. The load acting at the interior joints of each of the two bridge trusses is $16 \times 36 \times 0.16/2 = 46.0$ k, and half of this goes directly into the reactions (bridge piers) at 1 and 17.

Since all loads are parallel, one equilibrium equation, $\Sigma F_x = 0$, is trivial. Of the remaining two equilibrium equations, $\Sigma M_1 = 0$ yields the right reaction $R_{17} = 161$ k which, in this case, is equal to half of the total load

$$R_{17} = R_1 = 7 \times 46.0/2 = 161 \text{ k}$$

because of symmetry.

Considering that only a few forces are desired, the method of sections is convenient. In the present case, a single section will go through all three members, as shown in Figure 6.12*b*. We assume that tensile forces are positive and the directions of the unknown forces are drawn correspondingly. We recall that the force in a truss bar is directed along the bar.

There are three unknown forces acting on this free body which can be calculated from the three equations of equilibrium. If one writes these equations in a certain order, each unknown will occur in only one of the equations.

To evaluate $F_{5,7}$, we take moments about the intersection of the other two forces, hence

$$\Sigma M_6 = 48\ (161) - 32\ (46) - 16\ (46) - 24F_{5,7} = 0$$

from which $F_{5,7} = 230$ k (tension).

It is not necessary to carry decimals for values greater than 100; in fact, slide rule accuracy of about 1% is sufficient in truss analysis considering the uncertainties in loading and the approximations regarding joint fixities. Of course, the geometry must be calculated with greater accuracy for actual fabrication of the structure.

In a similar fashion, summation of moments about joint 5 (intersection of $F_{5,6}$ and $F_{5,7}$) yields $F_{4,6}$ directly:

$$\Sigma M_5 = 32\ (161) - 16\ (46) + 19.4\, F_{4,6} = 0$$

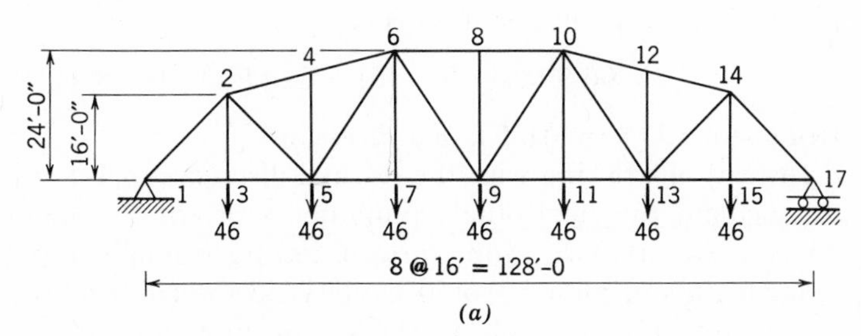


**Fig. 6.12**
Analysis of a Warren truss by the method of sections.

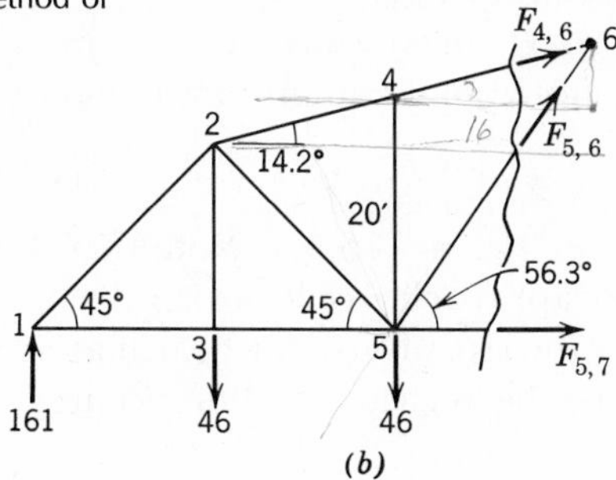

**Fig. 6.13**
Small highway bridge truss.

where 19.4 ft is the perpendicular distance from joint 5 to the top chord, 20(cos 14.2). Thus, $F_{4,6} = -228$ k (compression).

We can follow the same approach and calculate the force in the diagonal 5-6 by taking moments about the intersection of the top chord and the bottom chord. This occurs at 80 ft to the left of joint 5, since the slope of the top chord 2-6 is 1:4. The normal distance from that point to the extension of 5-6 is 80 (sin 56.3) = 66.5 ft.

The final moment equation is

$$-66.5\, F_{5,6} + 80\ (46) + 64\ (46) - 48\ (161) = 0$$

from which $F_{5,6} = -16.5$ k (compression).

Instead of working with the normal distances to 4-6 and 5-6, it is easier to calculate the horizontal components of these unknown forces first. Thus, if we break $F_{4,6}$ into vertical and horizontal components at joint 4, moments about joint 5 would easily yield the latter which, divided by cos 14.2 results in $F_{4,6}$. Similarly, the sum of moments about 4 gives the horizontal component of $F_{5,6}$.

$F_{5,6}$ could have been calculated from other equilibrium conditions; for example, horizontal equilibrium demands

$$F_{5,7} + F_{5,6}\ (\cos 56.3) + F_{4,6}\ (\cos 14.2) = 0$$

which also gives $F_{5,6} = -16.5$ k. Note that if both chords were horizontal, only the latter approach would work.

In practice, the truss is usually drawn accurately to scale and the normal distances could be measured. The accuracy of a careful graphical construction is almost the same (2 to 4%) as that obtainable by slide rule and frequently it is faster. We shall study a graphical method for getting the truss forces in the next section.

***Example 6.6*** Calculate the forces in several of the members of the same truss using the method of joints.

The reactions are computed first as in the previous example. We then examine the equilibrium of each of the joints in turn. We take joints that have not more than two unknown forces since only the equations $\Sigma F_x = 0$ and $\Sigma F_y = 0$ are available because of our assumption that all members at the joint meet at a point. At the start only joint 1 qualifies. Vertical and horizontal equilibrium conditions at joint 1 are

$$F_{1,2}(\sin 45) + 161 = 0; \qquad F_{1,2} = -228 \text{ k}$$

and

$$F_{1,2}(\cos 45) + F_{1,3} = 0; \qquad F_{1,3} = 161 \text{ k}$$

Next, we must go to joint 3 which has two unknown forces acting on it, as shown in Figure 6.14*a*. In this case, we have two pairs of opposing forces, thus $F_{3,2} = 46$ k and $F_{3,5} = F_{3,1} = 161$ k.

At joint 2, summation of forces in the *x* and *y* directions would result in

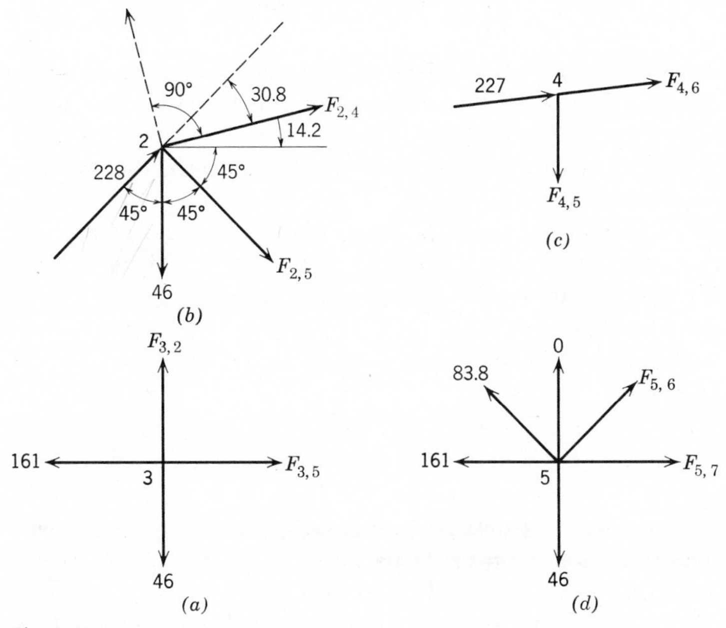


**Fig. 6.14**
Analysis of the truss of Fig. 6.12*a* by the method of joints.

two simultaneous equations; instead, we can consider equilibrium in directions normal to each of the unknown forces. Thus the summation along the line normal to 2-4 (Figure 6.14*b*) gives

$$228\,(\cos 59.2) - 46\,(\cos 14.2) - F_{2,5}\,(\cos 30.8) = 0$$

from which $F_{2,5} = 83.8$ k. Similarly, equilibrium perpendicular to $F_{2,5}$ would give $F_{2,4}$ without using the value of $F_{2,5}$, or alternatively, the horizontal equilibrium equation is

$$228\,(\cos 45) + F_{2,4}\,(\cos 14.2) + F_{2,5}\,(\cos 45) = 0$$

and therefore $F_{2,4} = -227$ k.

Next we consider joint 4 and see that only $F_{4,5}$ has a component normal to the top chord (Figure 6.14*c*) and therefore $F_{4,5} = 0$. Obviously, $F_{4,6} = F_{2,4} = -227$ k, which agrees closely with the value of $-228$ calculated by the method of sections in the previous example. In general, if two of three bars meeting at an unloaded joint are collinear, the force in the third bar is zero.

Having determined $F_{2,5}$ and $F_{4,5}$, only two unknown bar forces act on joint 5 (Figure 6.14*d*). The vertical equilibrium equation gives the force in the diagonal directly: $F_{5,6} = -16.2$ k, and horizontal equilibrium gives $F_{5,7} = 231$ k, which checks the value obtained previously.

Since the forces are determined from previously calculated member forces, an error carries through the analysis; for this reason it is wise to make at least one independent check using the method of sections. Modern desk calculators and electronic computers are well suited for the analysis of trusses by the method of joints.

The method of joints can easily be extended to three-dimensional trusses, except we then have three equations at each joint. Three-legged TV towers are examples of simple space trusses. A complex example is shown in Figure 1.7*h*. The geometry of space trusses is usually very complex but the basic ideas involved in their analysis are the same as for two-dimensional trusses.

## Arches

Arches are also axial-force structures in the sense that only compression exists in an arch under a certain loading. Thus, if we turn the cable of Figure 6.10 upside down, we obtain a pure compression structure. Arches are built of masonry, concrete, or steel to resist the compression and to resist buckling. Obviously, the shape of an arch does not change with the loading to avoid bending as in cables and therefore some bending may occur.

There are three types of arches depending on the support conditions: three-hinged, two-hinged, and hingeless (fixed) arches (Figure 6.15). Only three-hinged arches are statically determinate. The condition of zero moment at the internal hinge provides the fourth equilibrium equation for the calculation of the four reaction components.

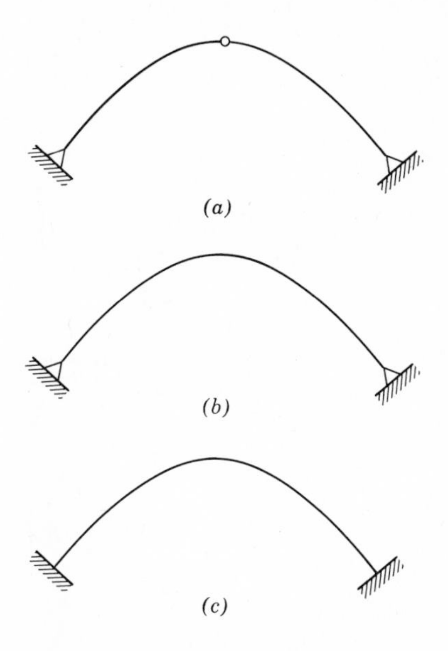


**Fig. 6.15**
Arch types. (*a*) Three-hinged. (*b*) Two-hinged. (*c*) Hingeless.

## 6.4 GRAPHICAL ANALYSIS

Numerous interesting graphical methods have been developed by structural engineers for the determination of forces in structures. We shall review a few of the most common procedures in this section. These are based on two facts: if only three nonparallel forces act on a body, they must go through a common point, and if the magnitudes of two forces acting on a free body are the only unknowns, the completion of a force diagram will determine their magnitudes.

***Example 6.7*** Evaluate the reactions of the three-hinged parabolic arch of Figure 6.16*a* under a concentrated load.

Since only three forces act on the structure ($R_1$, $R_4$, and $P$), a simple graphical solution is possible. We have seen in connection with the analysis of trusses that if a pin-ended member is not loaded, its end forces must be

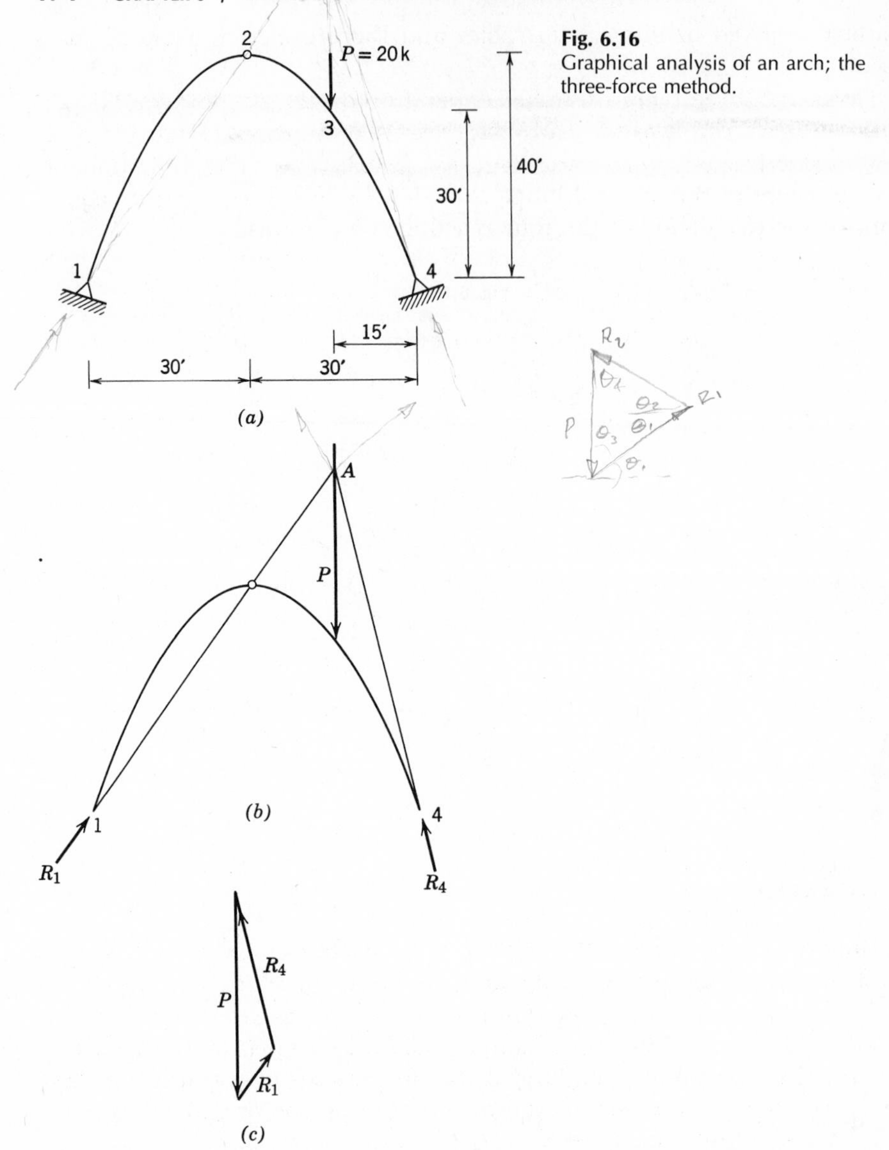


**Fig. 6.16**
Graphical analysis of an arch; the three-force method.

collinear. Similarly, in the case of this arch, since member 1-2 is pin-ended and unloaded, the reaction at 1 must go through the joint at 2. Only in this way can the bending moment at 2 be zero. The extension of this line meets the line of the load at *A* (Figure 6.16*b*) and thus $R_4$ must also pass through *A*. Knowing the directions of $R_1$ and $R_4$, we can construct the force diagram of Figure 6.16*c* to an appropriate scale.

**Fig. 6.17**
Three-hinged Gothic arch.

*Unadilla Laminated Products*

If several forces were acting on this arch, this construction would not work directly, but we could evaluate the reactions for each of the loads in turn and use the principle of superposition (see Section 5.2) to obtain the total solution.

Several other types of structures can be analyzed by the graphical method. In the case of trusses the directions of all forces are known. We examine free bodies of joints that have no more than two unknown forces acting on them, as in the method of joints. Completion of the force diagram yields the magnitude of the unknown forces. The force diagram for the joint must form a continuous, closed polygon, otherwise the unbalanced force components would violate equilibrium.

We may construct a force diagram for each joint of a truss. This is shown in Figure 6.18*a* for a simple support for a cantilevered roof over an entrance. We follow the same order as in the case of the method of joints. Since each joint equilibrium equation contains at least one previously determined bar force, it is possible to combine these force polygons as shown in Figure 6.18*c*. This is called a *Maxwell diagram.* It is made of polygons, each of which reflects the equilibrium of a joint. At each joint we start the polygon with all the known forces and proceed around the joint in the same sense, clockwise in the present example. It is important to watch the sense of each of the force vectors in the diagram; it is directed

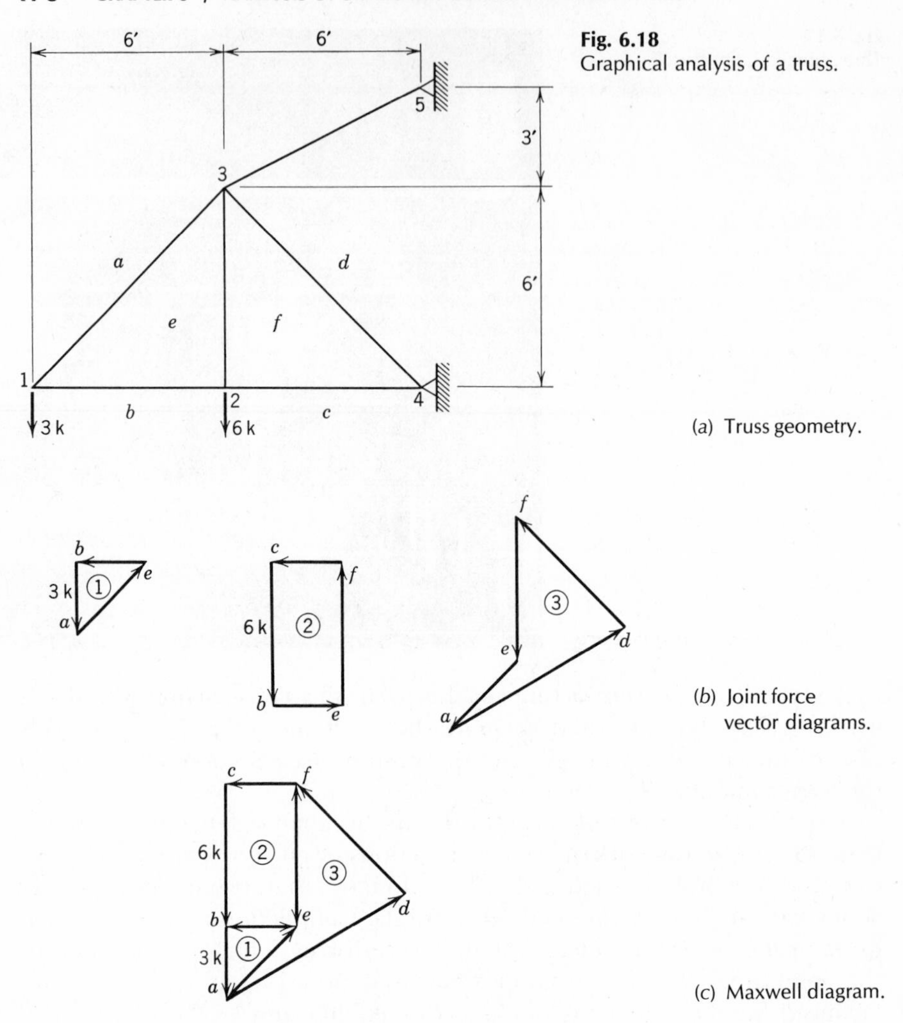


**Fig. 6.18**
Graphical analysis of a truss.

(a) Truss geometry.

(b) Joint force vector diagrams.

(c) Maxwell diagram.

away from the joint if the force is tensile and toward the joint if the bar force is compressive. Therefore, the arrow on a vector is changed when it is considered the second time, that is, at another joint, in the Maxwell diagram. The numbers inside the force polygons in Figure 6.18 indicate the joint number for which the polygon expresses equilibrium. The vector between two such numbers represents the member force between the respective joints. When some polygons overlap, it is more convenient to use another notation. We use lowercase letters to identify the fields between applied forces and truss bars (Figure 6.18*a*). These symbols can then be used to identify the ends of the forces in the force polygons (Figure 6.18*b* and 6.18$_c$). The signs of the bar forces are determined by proceeding clockwise around each joint (d-f-e-a on joint 3) and noting the corresponding vector direction in the force polygon (d-f pushes, f-e pulls, etc., on joint 3).

The Maxwell diagram may appear complicated if more than three forces meet at some joints or when a number of force polygons overlap. This is illustrated by the graphical solution of the truss of Figure 6.12, as indicated in Figure 6.19 (only half of the symmetrical truss and loading is analyzed).

Graphical methods are very fast and their accuracy is almost the same as obtainable with a slide rule. Graphical statics was especially popular before modern desk calculators and computers made numerical analysis easier. Yet, many engineers use the almost foolproof graphical methods even today.

The graphical approach is well suited for the determination of the forces in cables loaded with concentrated loads. For example, consider the cable of Figure 6.7*a*. Assume that the vertical component of the left reaction was calculated: $V_1 = 55.0$ k from Section 6.3. Again, we need to know the elevation of one point on the cable, say that $y_2 = 25.0$ ft is given.

Since the force in cable segment 1-2 opposes the reaction, the vertical component of $F_{1,2}$ is 55.0 k. Thus, we can determine $F_{1,2}$ since its direc-

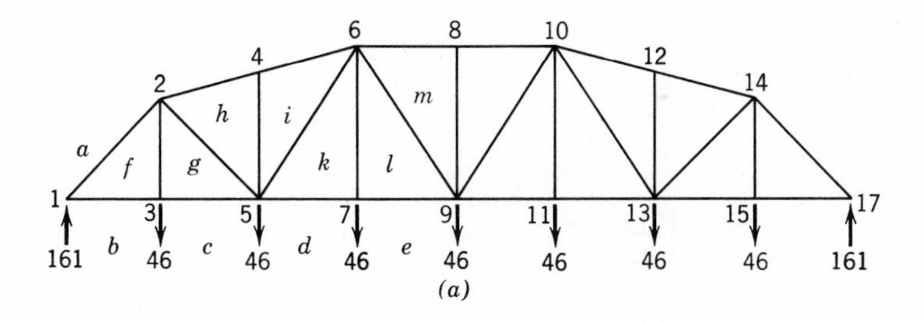

(*a*)

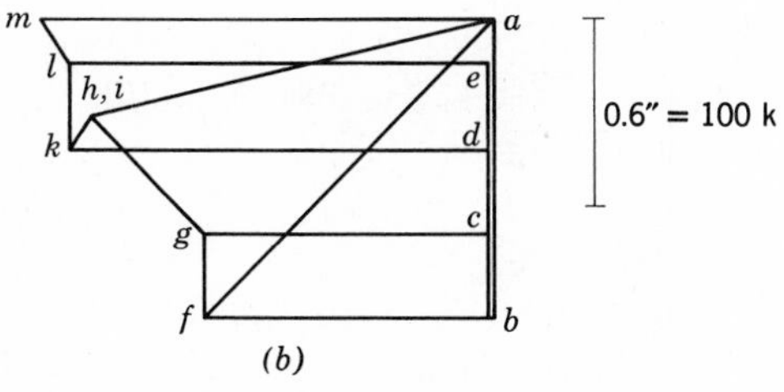

(*b*)

**Fig. 6.19**
Graphical solution of the truss of Fig. 6.12.

tion and its vertical components are known. Figure 6.20*a* shows the determination of $F_{1,2}$. We can construct the direction of each subsequent cable segment and the force therein from the direction of the previous cable segment, its force, and the load between the two cable segments. This is done in Figure 6.20*b* in such a way that the force parallelogram is under the respective loads and the cable forces are continuous. The line of the cable forces is called a *string polygon* or a *funicular polygon;* we shall see that it is also very useful in the solution of other types of problems.

Again we combine force triangles into a force polygon, drawn to an arbitrary force scale (Figure 6.20*c*). The cable forces meet at a common point, called the *pole* of the force diagram. The lines going through the pole are called *rays* of the force polygon. Obviously, the last cable segment (last ray) defines the right reaction. Each triangle in the force polygon sig-

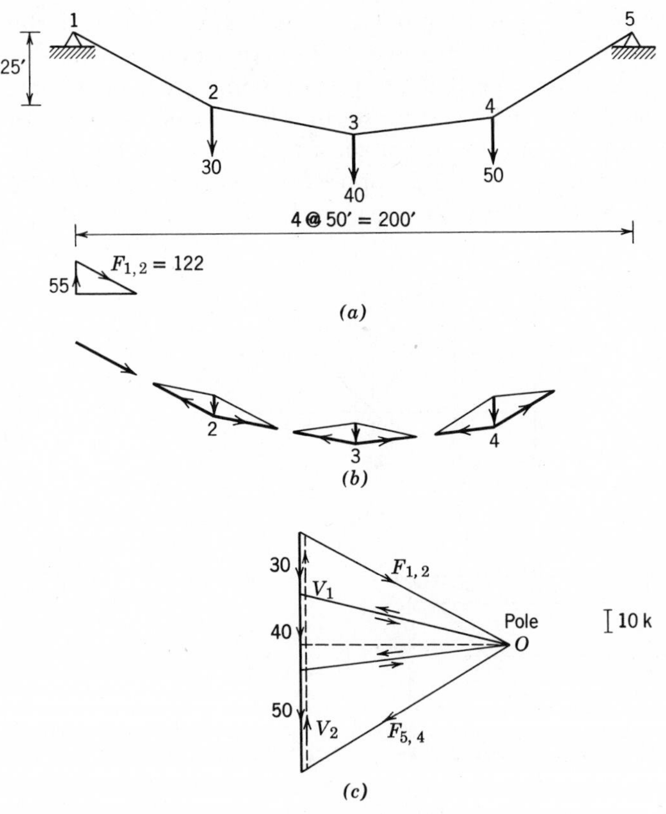


**Fig. 6.20**
Graphical solution of a cable; the string polygon.

nifies the equilibrium of a point and this is marked by a corresponding number in the triangles. The vectors of the forces $P_2$, $P_3$, $P_4$, and the closing rays $R_1 = -F_{1,2}$ and $R_5 = -F_{5,4}$ close in a loop since this polygon represents the overall equilibrium of the structure. The intermediate rays are each made up of two equal and opposing forces which cancel out when we superpose the triangles, as indicated on the diagram. The horizontal projection of the rays, which is the distance between the pole and the line of the loads, is the common horizontal component of the cable forces and the reactions. The horizontal line through the pole subdivides the line of the force vectors into two forces, the vertical components $V_1$ and $V_5$ of the reactions.

We can use the string polygon to determine the reactions as well. If we did not know $V_1$ to begin with, then we would not know what force vector would be measured along the first ray in the force polygon. In such a case, we arbitrarily assume the position of the pole, $O'$ along the first ray (Figure 6.21*b*) and complete the string polygon and the force polygon as before. The last line does not go through the right reaction, point 5. The closing line of this string polygon 1-5′ again determines the ray in the force polygon that cuts out the vertical components of the reactions (point $E$ in Figure 6.21*b*). We know that the closing line of the string polygon must be horizontal in this case and thus we draw a horizontal ray from point $E$; its intersection with the known direction of the first ray is the new pole $O$.

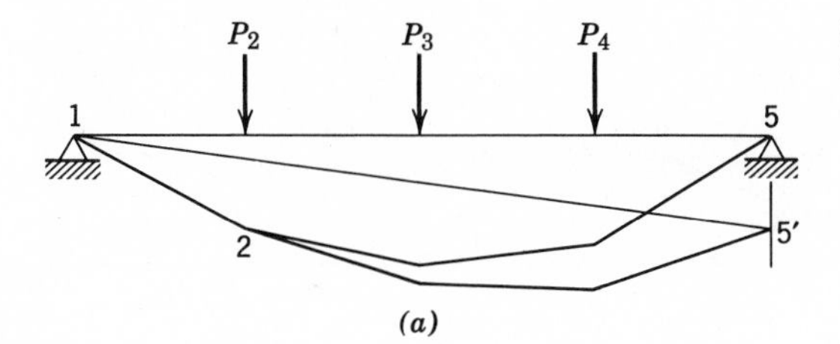


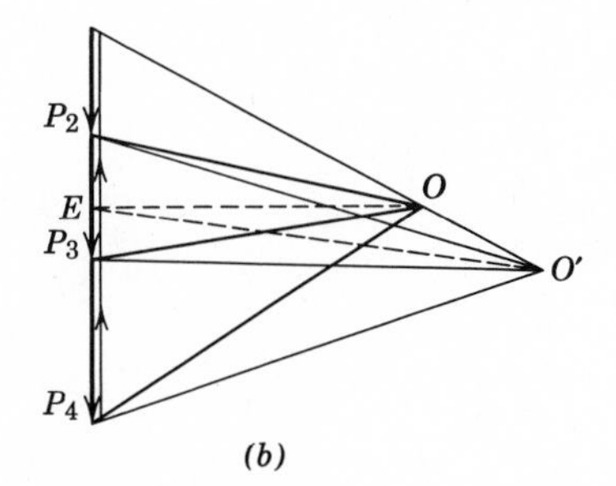


**Fig. 6.21**
Graphical determination of reactions.

Now we can draw new funicular and force polygons, also shown in Figure 6.21, which give us the magnitude and the direction of the reactions (and the cable forces in the present example).

The above method can be extended to obtain the solution for the case where the sag of an intermediate point is given. Any funicular polygon that goes through the end points is a cable in equilibrium and we select one by specifying a sag or a force.

***Example 6.8*** As another illustration of the graphical approach, let us obtain the reactions of the beam shown in Figure 6.22.

We know that reaction 1 must go through point 1 and that reaction 5 must be vertical. Proceeding in the same manner as in the case of cables, we draw the force diagram *ABCD* and arbitrarily select a pole. Then the funicular polygon can be constructed starting at point 1, Figure 6.22*b*. The final line

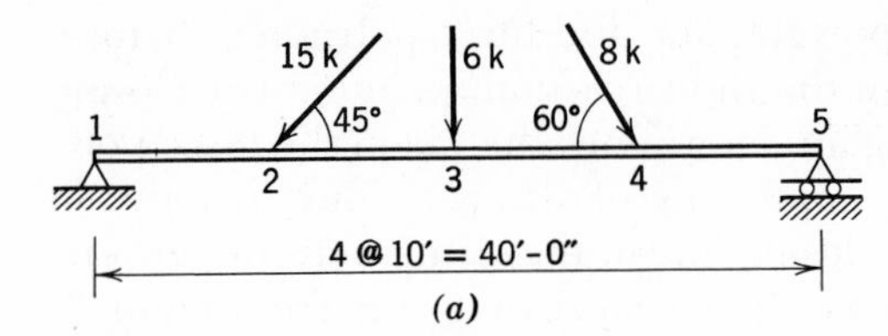


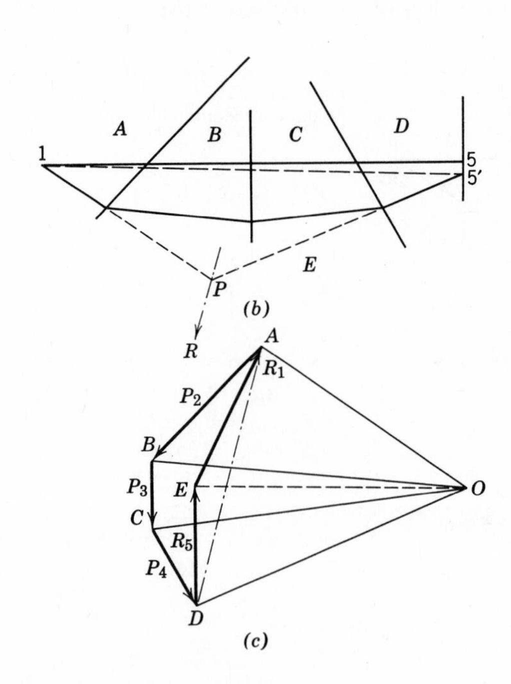


**Fig. 6.22**
Graphical determination of beam reactions. (*a*) Loaded beam. (*b*) Funicular (string) polygon. (*c*) Force polygon.

intersects the line of the right reaction at point 5′ instead of at 5. The closing line of the string polygon is 1-5′ and the ray parallel to it locates point *E*, giving the two reactions since $R_5$ follows $P_4$ and is vertical. The force vector *E-A* which completes the polygon is the left reaction $R_1$. Note that both the string polygon and the force polygon are closed. The string polygon does not go through point 5 but since we were seeking the reactions of the beam, the actual position of the string polygon is immaterial. We thus see that the string polygon is an aid in obtaining reactions and is not restricted to the analysis of cables. This approach also works when the loads are parallel, whereas the three-force method would not work because the resultant of the loads would not intersect either of the reactions.

We can use the construction shown in Figure 6.22 to find the resultant of forces. The direction and the magnitude of the resultant of the loads $P_2$, $P_3$, $P_4$ is given by *A-D* in the force polygon, but we do not know where it acts relative to the forces. Each of the forces can be replaced by two rays in the force polygon but, as we have already observed in connection with Figure 6.21, the pairs of intermediate rays cancel out. This means that the forces represented by the first and the last rays (*OA* and *OD*) are statically equivalent to the three loads. The resultant *R* of the loads must therefore pass through the intersection *P* of the first and last lines of the string polygon, as indicated in Figure 6.22*b*.

## 6.5 BENDING STRUCTURES: BEAMS AND FRAMES

The majority of common structures have numerous flexural elements that carry the load primarily by transverse bending rather than by axial action. We have seen the features of such structures in Chapter 4. Beams in buildings and bridges provide the best example of flexural members. The vertical and horizontal loads on buildings cause significant bending moments in beams and also in columns. Since the bending stresses are relatively large in slender members, the design of beams and frames depends strongly on the determination of the moments along each member. The variation of the moment in bending structures influences their behavior and it is essential to develop a clear understanding of this behavior. It is customary to draw diagrams showing the moment, shear, and sometimes the axial load variation along the members.

***Example 6.9*** Determine the variation of the moment and shear along the roadway members of the Solleks River Bridge under the loading shown in Figure 6.4. The reactions are reproduced in Figure 6.23*a* from Figure 6.5. The deflected shape of the structure is sketched in Figure 6.23*b*.

First we plot the shear *V* in the beam on a diagram that indicates the centerline of the member. We shall adopt the convention that a shear force is positive if it acts upward on the left end of a member. Thus the shear at the left end of the bridge is negative. We measure 1.7 k downward from a reference axis in Figure 6.23*c*. Since no vertical force components act between points 1 and 2, the shear does not change in that segment. At point 2 the shear changes by +17.8 k, the value of the vertical component of the reaction there. The rest of the *shear diagram* is constructed in a similar fashion. The change in shear at the right end, −0.6 k, brings the value down to zero. This is a check on the vertical equilibrium condition. Note that no external forces act at the internal pins at 3 and 4 and correspondingly there is no change in the shear there.

Next we plot the bending moment $M$ on a diagram similar to the shear diagram. We define moment as positive if it causes tension on the bottom

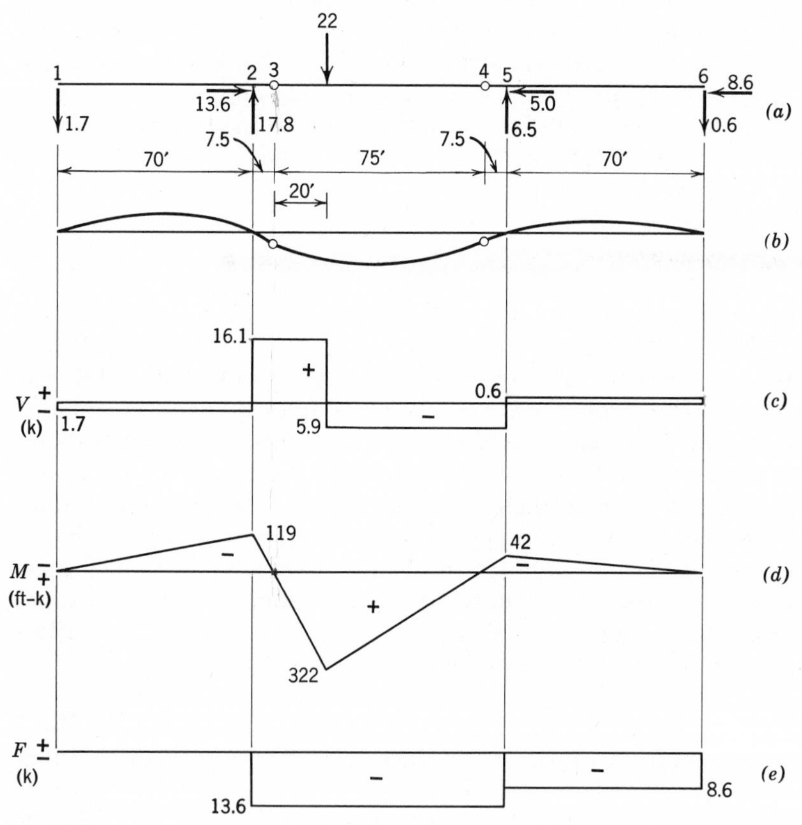


**Fig. 6.23**
Shear, moment, and axial force diagrams for the Solleks River Bridge.

face of the beam and we plot positive moments below the reference line, corresponding to the *tension side* of the member, for reasons that will become clear subsequently.

The moment is zero at 1 and it increases linearly between 1 and 2, being equal to $-1.7x$ where $x$ is measured from 1. This is obtained by cutting member 1-2 at a distance $x$ from the left end and solving for the unknown internal moment at the cut. Hence $M_2 = -1.7 \times 70.0 = -119$ ft-k is plotted above 2. The moment between 2 and 3 again depends linearly on the distance between the forces and the points considered; in fact, the *moment diagram* must clearly be linear in segments where there are no transverse loads acting. We could calculate $M_3$ by taking moments about 3 for either free body or we can simply continue the line through 3 since the moment must be zero at the internal hinge. The entire moment diagram is drawn in Figure 6.23*d*. Note that the positions of zero moment and the internal hinges coincide.

It is instructive to compare the moment diagram and the deflected shape. The curvature of the elastic curve should correspond to the moment diagram. At inflection points, where the curvature changes sign, the moment must be zero. This relationship between the elastic curve and the moment diagram is useful in sketching either diagram and some engineers draw these diagrams simultaneously (see Example 6.10).

The *axial force diagram* is usually quite trivial and therefore it is seldom drawn. It is shown in Figure 6.23*e* using a convention that tension forces are positive.

Obviously, several sets of such diagrams corresponding to various loading conditions must be considered in design. The three segments of the bridge would be designed for maximum values of shears, moments, and axial loads, which can be picked off these diagrams.

The sign convention for moment and shear for beams is illustrated in Figure 6.24. Note that positive shear forms a clockwise couple on a segment.

To aid in the understanding of the action of structures, you should

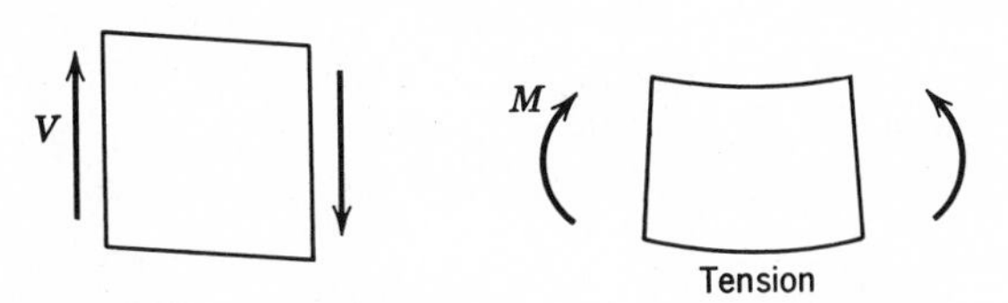


**Fig. 6.24**
Sign convention for positive shear and moment in a beam.

learn to sketch the $M$ and $V$ diagrams and the deflected shape without any calculations. We shall practice this with the help of several examples. Following Example 6.11, we shall discuss relationships between load, shear, moment, deflection, and slope, which will facilitate quantitative development of such diagrams.

***Example 6.10*** Sketch the deflected shape and the $M$ and $V$ diagrams for the Solleks River Bridge for the loading shown in Figure 6.25*a*.

The deflected shape or elastic curve is sketched in Figure 6.25*b*. The approximate positions of the inflection points, where $M = 0$ and the curvature changes sign, are marked with solid circles. We also indicate the tension side by the marking "*t*".

The shape of the moment diagram for a simply supported beam under a uniformly distributed load is a parabola, whereas it is composed of linear segments under concentrated loading. If both distributed and concentrated loads act, the moment diagram is parabolic under the distributed load but has a local peak or cusp at the concentrated load. For the sake of illustration, the three simple-beam moment diagrams are plotted in Figure 6.25*c*, neglecting the internal hinges. Note that the shape of these diagrams resembles the shape of a flexible cable under the same loading.

In the continuous beam, the moment is not zero at internal supports. In the present case (and in most practical situations), the tension side is on top of the beam at these supports (Figure 6.25*b*) and correspondingly we plot the moment above the axis ($M_2$ and $M_3$ in Figure 6.25*d*). These are negative moments. The final moment diagram is the superposition of the simple-beam diagrams and the negative moment line due to the internal support forces. This superposition can be done two ways. In one method, the simple-beam moments are simply measured from the plot of negative moments as a base line (Figure 6.25*d*). The final moment diagram is shown shaded. The construction indicates that the simple-beam moments are reduced because of the continuity at the supports. The moment must be zero (cross the base line) at the internal hinges. Note that the moment diagram is on the tension side of the beam. In the second approach, the negative moment diagram is subtracted directly from the positive areas; this construction is simple but the values of the moments are not measured from a horizontal line (Figure 6.25*e*). Ordinarily, it is not necessary to plot the simple-beam moments first; we have done it here to point out their superposition with the negative support moment line.

The shear diagram is plotted in Figure 6.25*f*. Proceeding from left to right we simply plot the forces in the direction they act. The shear diagram is a straight sloping line under uniformly distributed loads. The shear is

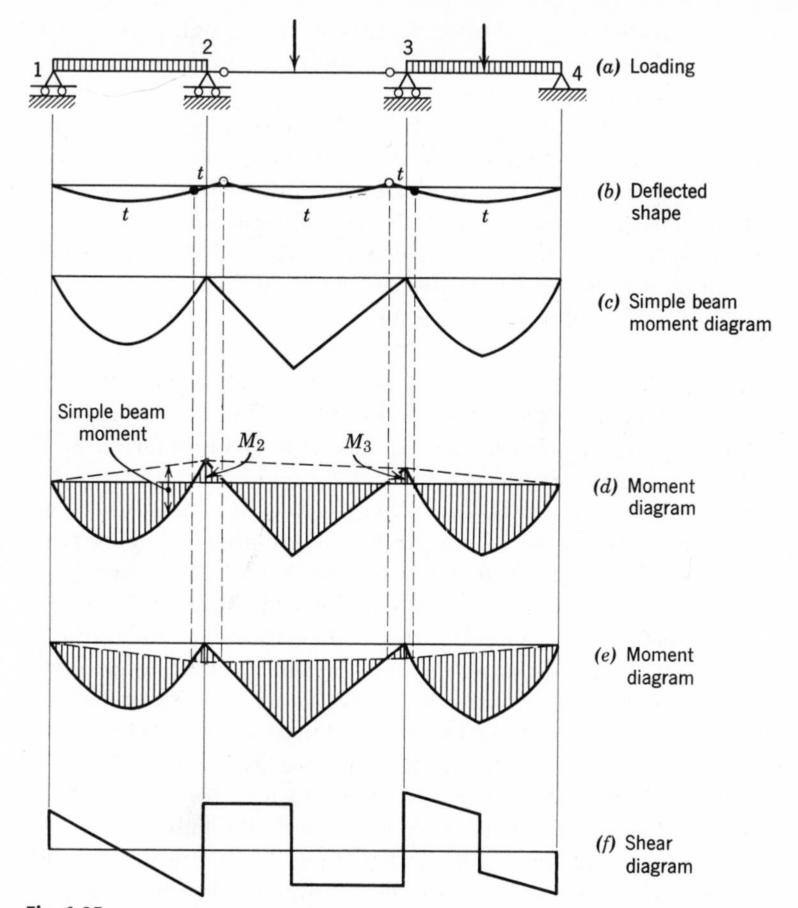


**Fig. 6.25**
Sketching of shear and moment diagrams.

constant over segments where no load acts and the moment diagram is linear there; this follows from the fact that the moment is a linear function of distance (see also the discussion in the next subsection). It is customary to draw the shear diagram first before the moment diagram, but in this example we have placed emphasis on the moment diagram by considering it first.

The importance of gaining complete understanding and skill in drawing these diagrams cannot be overemphasized. The distribution of stresses, calculation of deflections, the placement of the reinforcement in concrete members, and other situations in structural design are based on the knowledge of the shear and moment diagrams. The role of creativity, architectural factors, and the presence of uncertainties in the design process are partially qualitative, but there cannot be any doubt or questions about the correct moment diagram.

There are a number of problems for drawing $V$ and $M$ diagrams at the end of this chapter and you should invent additional problems to obtain sufficient experience. Some of the structures are indeterminate but it is instructive to sketch qualitative $M$ and $V$ diagrams for such structures at this stage; Chapter 7 explores the subject further. The drawing of bending moment diagrams for frames involves a few special considerations that are illustrated in the following example.

***Example 6.11*** Sketch the deflected shape and the $M$ and $V$ diagrams for the gable frame of Figure 6.26*a*. This type of frame is often used in small industrial buildings and churches. Since it has a third hinge at point 4, it is a statically determinate structure.

When sketching deflected shapes, it is helpful to pay attention to the tension side and the curvature of the deflected members. The sloping members 2-4 and 4-5 bend downward as beams and therefore both columns bend outward, accompanied by rotations of the joints 2 and 5. Obviously, the entire outside faces of the two columns are in tension. The outside of each rigid joint is also in tension but the middle portions of the sloping members bend downward. Thus, there must be inflection points in the outer halves of these members. These are marked on the elastic curve in Figure 6.26*b*. The deflections are greatly exaggerated in the sketch; actually the deflections are of the order of a fraction of a percent of the span. The tension sides are noted with the letter *t*.

Since the columns try to move outward at their bases, the horizontal components of the reactions must be directed inward. The shear is constant in the columns (Figure 6.26*c*), negative in 1-2 and positive in 5-6. The distribution of shear in the sloping members is similar to that in a beam. The end shears at the ends of the members (points 2, 4, and 5) are produced by the appropriate force components in the adjoining members. Specifically, the pin at 4 transmits a force that causes end shears and axial forces in members 4-2 and 4-5.

The moments in the two columns start with zero at the hinged bases and increase linearly due to the effect of the horizontal components of the reactions. The distribution of moments in the sloping members is similar to the moments in equivalent beams. At point 4, the moment is zero. At the rigid joints (points 2 and 5), the ends of these members cannot rotate freely because of the bending stiffness of the columns. The end fixity is between that of a hinged and a fixed end of a beam. The moment at the lower end 2 of the sloping member 2-4 must be equal to the moment

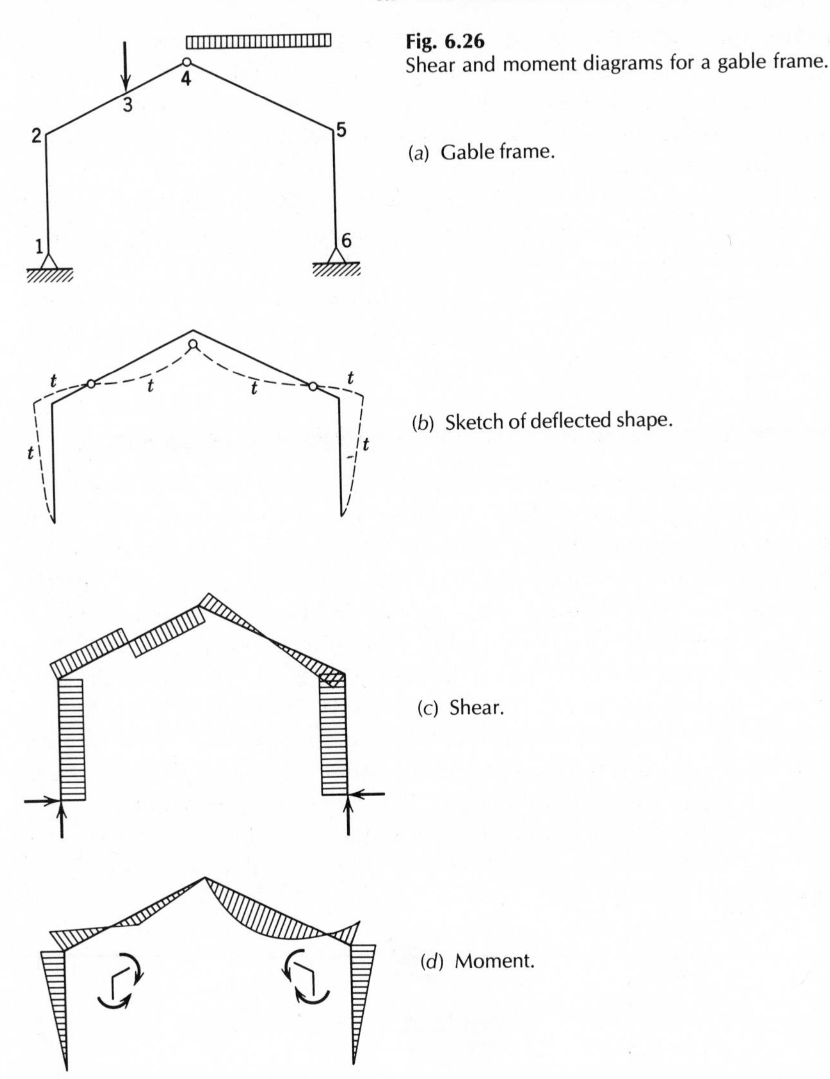


**Fig. 6.26**
Shear and moment diagrams for a gable frame.

(*a*) Gable frame.

(*b*) Sketch of deflected shape.

(*c*) Shear.

(*d*) Moment.

on the top of the column 1-2. Thus, we plot the moment on the tension (outer) side of the columns, carrying the moment around the corner at joints 2 and 5 (Figure 6.26*d*). At a rigid connection of two members, the end moments are equal, as shown in the small diagrams of Figure 6.26*d*, where the moments acting *on the joints* are indicated. These moments are always plotted on the same side of the structure. If the angle of the rigid joint "tries" to decrease, the tension is on the outside, otherwise it is on the inside, although we emphasize that the angle remains constant at such joints. The moment changes sign in the sloping members; the points of zero moment must coincide with the inflection points noted in Figure 6.26*b*. The distribution of moments is parabolic in member 4-5 because of the uniformly distributed loading.

When more than two members are joined rigidly, the algebraic sum of the member end moments and external moments applied to the joint must be equal to zero and should be plotted accordingly.

Having completed the sketching of the $M$ diagram, we should compare it with the deflected shape and revise either or both diagrams to match the points of zero moment and inflection points.

## Relationships Between Load, Shear, Moment, Deflection, and Slope

The moment and shear in a beam and the deflection of a beam are related to the loading in an exact manner which can be expressed with equations. However, you should continue to sketch the $V$ and $M$ diagrams as presented in the previous section until you can predict the shapes with confidence. The relationships to be derived in this section are useful both in sketching these diagrams and in obtaining numerical values.

Consider an element of a beam having length $dx$ (Figure 6.27). It is loaded by a distributed load $p$ which need not be constant over the length of the beam. The internal forces $V$ and $M$ on the left face are increased by $dV$ and $dM$ to obtain the corresponding values acting on the right face. This beam element is in equilibrium. The vertical equilibrium $\Sigma F_y = 0$ gives $pdx = -dV$ and the sum of moments about point $A$ gives $M - (M + dM) + Vdx - (pdx)dx/2 = 0$. If we neglect the higher order terms involving $(dx)^2$, we get

$$p = -\frac{dV}{dx} \tag{6.4}$$

$$V = \frac{dM}{dx} \tag{6.5}$$

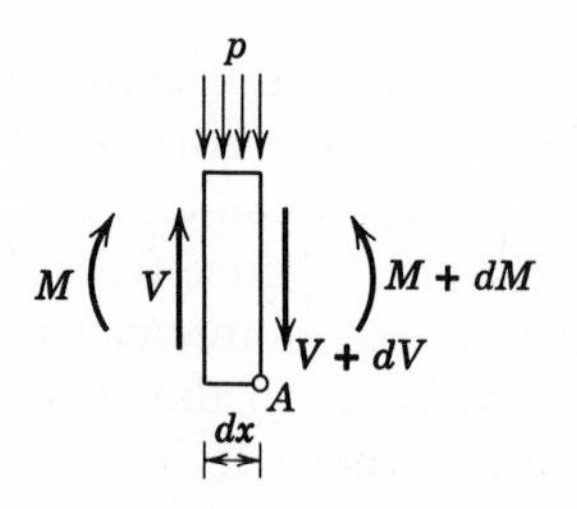


**Fig. 6.27**
Forces on a beam element.

These simple equations are useful in calculating the $V$ and $M$ values along flexural members as well as for helping determine the proper slopes and curvatures of the $V$ and $M$ diagrams. The first equation says that the rate of change of shear is equal to the value of the distributed load; hence, the shear is constant where $p = 0$, and the shear diagram is linear where $p$ is constant (see, e.g., Figure 6.25 and 6.26). This equation is not usable at concentrated loads that represent discontinuity in the $V$ diagram. In some books the negative sign is avoided by defining positive loads upward. The change of shear from point 1 to point 2 is

$$\Delta V = -\int_1^2 p dx \tag{6.6}$$

According to Equation 6.5 the rate of change of the moment is equal to the intensity of shear. Thus, if $V = 0$ over a region, $M$ is constant there. For constant $p$, $V$ is linear and $M$ has quadratic variation. Furthermore, the change in moment over a distance $dx$ is equal to $dM = Vdx$. The change in moment from point 1 to point 2 along a segment is

$$\Delta M = \int_1^2 V dx \tag{6.7}$$

The maximum value of the moment in a flexural member is of prime interest. According to calculus, local maxima occur where the derivative is zero or where the function is discontinuous. From Equation 6.5, we see that $dM/dx = 0$ occurs where $V = 0$. Thus, we come to a very important conclusion: the moment reaches local maximum values wherever the shear is zero, that is, where the $V$ diagram crosses the axis. At a concentrated load or a reaction where the shear diagram changes sign, the shear is assumed to pass through zero value over a very small distance and thus maximum moment occurs there.

Maximum moment may also occur where the moment diagram is discontinuous, even though the shear is not zero. This happens where concentrated moment acts, either as applied load or as a fixed-end reaction, for example at the fixed end of a cantilever beam. The local maximum moment can be positive or negative.

***Example 6.12*** Calculate the maximum moment in member 1-3 of the Solleks River Bridge (Figure 6.4*a*) as a result of the self-weight of the members. The cross-sectional area of the girders is 3.5 ft$^2$ (see Figure 1.6*a*) and therefore the load is $3.5 \times 120 = 420$ lb/ft $= 0.42$ k/ft, where 120 lb/ft$^3$ is the unit weight of the light-weight concrete (Table 3.1). The force transmitted from the central span 3-4 is $75.0 \times 0.42/2 = 15.7$ k. Thus we

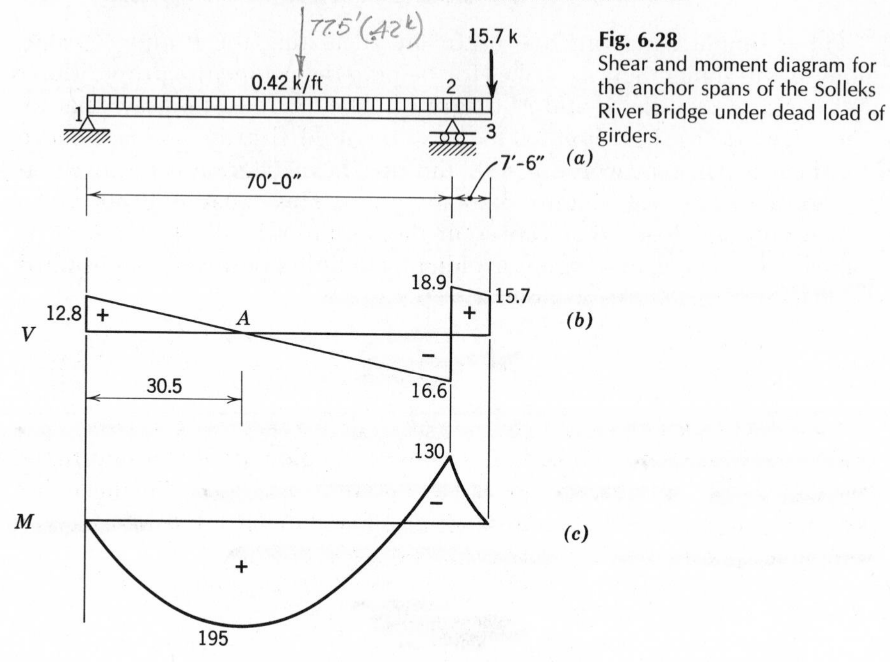


**Fig. 6.28**
Shear and moment diagram for the anchor spans of the Solleks River Bridge under dead load of girders.

arrive at the loading shown in Figure 6.28*a*. The axial forces are not considered here.

The right reaction is calculated from $\Sigma M_1 = 77.5(0.42)77.5/2 + (77.5)(15.7) - 70.0R_2 = 0$, from which $R_2 = 35.5$ k. Note that in this summation the moment of the resultant of the entire distributed load was taken. From vertical equilibrium $R_1 = 77.5\ (0.42) + 15.7 - R_2 = 12.8$ k. The shear diagram is shown in Figure 6.28*b* and can be completed by calculating the integral of the load (Equation 6.6). It crosses the zero axis at points $A$ and 2, the former is at $12.8/0.42 = 30.5$ ft from point 1. The maximum moment must occur at these points. $M_A = 12.8 \times 30.5/2 = 195$ ft-k (which is the area of the shear diagram between 1 and $A$). This is the maximum positive moment, whereas $M_2 = (18.9 + 15.7)7.5/2 = 130$ ft-k is the maximum negative moment (tension on top). These calculations indicate the advantages of using Equation 6.7 in getting moments. The moment diagram (Figure 6.28*c*) is not tangent to the axis at 3 because of the concentrated load there, but otherwise (with only the uniform load) the moment diagram would approach the end of the cantilever or overhanging portion of the beam tangentially. This follows from the fact that the shear is then zero at the tip of the cantilever and $dM/dx = V = 0$ gives a tangent of zero slope.

The displacement or deflection $y(x)$ of a beam at its neutral axis depends on the moment and shear in the beam. We shall neglect deflection due to shear as it is very small in most beam-type structures. First, we derive the relationship between the moment and the curvature as a function of $y(x)$ to get an equation between the moment and the deflection.

Consider a beam segment under the action of positive moments (Figure 6.29*a*). The cross sections $dx$ apart undergo relative rotations $d\phi$ corresponding to a curvature $1/R$, where $R$ is the radius of curvature (the local deformed centerline is a circle with a center at point 0). This relative rotation is related to the linear bending strain distribution over the height of the beam (Figure 6.29*b*). From similar triangles

$$d\phi = \frac{dx}{R} = \frac{\epsilon dx}{c} \tag{6.8}$$

where $\epsilon$ is the strain and $c$ is the distance from the beam neutral axis to the fiber where $\epsilon$ occurs. The stress in the same fiber is $f = \epsilon E$ which, if caused by bending, is also equal to $f = Mc/I$. Equating these we get $\epsilon/c = M/EI$. Therefore, combining this equation with Equation 6.8, we get

$$\frac{1}{R} = \frac{M}{EI} \tag{6.9}$$

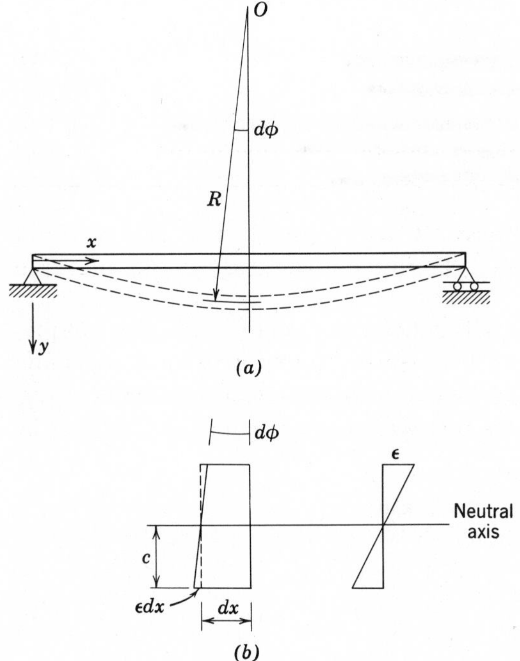


**Fig. 6.29**
Geometry of deformed beam element.

This is the basic moment-curvature relationship. According to this equation, positive moments cause positive curvature for which the center of curvature is above the beam. If the moment and the bending stiffness $EI$ are constant between two points, the deflected shape is a circular arc since then $R$ is also constant.

We are usually not interested in the curvature itself, but we can calculate $y(x)$ from the known curvature $1/R$. It is known from calculus that the curvature of a line $y(x)$ is given by

$$\frac{1}{R} = \frac{(d^2y/dx^2)}{[1 + (dy/dx)^2]^{3/2}} \tag{6.10}$$

Since we are concerned only with small deflections (say $y_{max} \leq L/100$), the square of the slope can be neglected in comparison with unity in the brackets. Hence, using Equations 6.9 and 6.10, we have

$$\frac{d^2y}{dx^2} = -\frac{1}{R} = -\frac{M}{EI} \tag{6.11}$$

where the minus sign was imposed because we have taken downward deflections positive and therefore for positive $M$ and increasing $x$ the slope $dy/dx$ should increase and $d^2y/dx^2$ must be negative (Figure 6.30).

Using Equations 6.4, 6.5, and 6.11, the following set of expressions can be assembled:

$$y(x) = \text{deflection of the member} \tag{6.12a}$$
$$dy/dx = \text{slope of the member} \tag{6.12b}$$
$$d^2y/dx^2 = -M/EI = \text{curvature of the member} \tag{6.12c}$$
$$d^3y/dx^3 = (d/dx)\,(d^2y/dx^2) = -V/EI \tag{6.12d}$$
$$d^4y/dx^4 = (d/dx)\,(-V/EI) = -p/EI \tag{6.12e}$$

where a constant bending stiffness $EI$ was assumed. The last equation presents a relationship between the distributed loading $p(x)$ and the deflection $y(x)$. Thus, we can evaluate the latter by integrating Equation 6.12$e$ four times. This method of deflection calculation is not simple, except in very elementary cases, because the four integration constants must be evaluated from the boundary conditions (see also Problem 6.32). More efficient deflection analysis methods are presented in Volume 2.

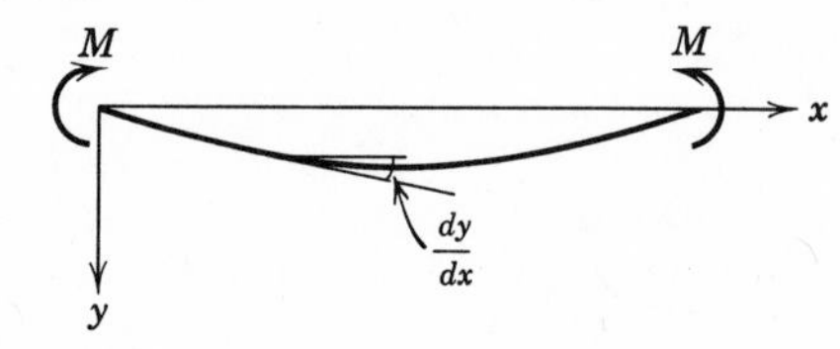


**Fig. 6.30**
Elastic curve of a beam under constant moment.

***Example 6.13*** Calculate the maximum deflection of the structure of Example 6.12.

The moment in the main span is $M(x) = R_1x - qx^2/2 = 12.8x - 0.21x^2$ and the curvature of the beam is $d^2y/dx^2 = -M/EI$, where $x$ is measured from the left support. Integrating twice, we get

$$EI\frac{dy}{dx} = -\frac{12.8}{2}x^2 + \frac{0.21}{3}x^3 + C_1$$

and

$$EIy = -\frac{12.8}{6}x^3 + \frac{0.21}{12}x^4 + C_1x + C_2$$

The boundary conditions are that $y = 0$ at $x = 0$ and at $x = 70.0$ ft. Thus, $C_2 = 0$ and $C_1 = 12.8\ (70)^2/6 - 0.21 \times 70.0^3/12 = 4450$. Therefore, the deflection of the beam is

$$y = (-\frac{12.8}{6}x^3 + \frac{0.21}{12}x^4 + 4450x)/EI$$

The maximum deflection occurs where $dy/dx = 0$, or

$$\frac{-12.8}{2}x^2 + \frac{0.21}{3}x^3 + 4450 = 0$$

The solution of this equation is $x = 33.0$ ft and the deflection there is

$$y_{max} = \frac{0.91 \times 10^5\ ft^3\text{-}k}{EI}$$

The deflection at the end of the cantilever can be calculated in a similar manner except the function $M(x)$ should include the effect of the right reaction. The moment of inertia of the bridge girder is $I = 164{,}300$ in$^4$ and using $E = 3.4 \times 10^3$ ksi, the maximum deflection of the bridge girder under its own weight is 0.28 in. This deflection may increase with time due to the creep of concrete. The actual deflection differs from the above value because of the camber (upward deflection) as a result of the prestressing of the girder, which was used to counteract the additional DL and the LL effects.

## 6.6 FORCES IN PLATE AND SHELL STRUCTURES

The structural action of surface structures is fundamentally different from that of structures made of linear elements. Some of the differences were explained in Chapter 4. This section presents the forces in surface structures in general terms; their detailed study is covered in advanced books on plate and shell theory.

Consider a flat plate (e.g., a table top or a floor slab) supported around its edges. A load on the plate is carried primarily by flexural action to the supports as in the case of beams. However, now we have two types of moments $M_x$ and $M_y$ and shears $V_x$ and $V_y$ acting along the edges of an element, as shown in Figure 6.31*a*. In addition, torsional moments $M_{xy}$ and $M_{yx}$ are produced by the load. The stresses in a plate supported on all edges are much smaller than in equivalent beam strips because of the two-way action of plates. The deflection of any point of the plate is the solution of a fourth-order partial differential equation. Frequently, numerical integration methods are used to obtain solutions that satisfy the given boundary conditions.

When the deflection of a plate-type structure is great or when it is not flat initially, such as in a curved shell element, membrane forces are created in addition to the bending forces. The membrane (or in-plane) forces are analogous to the axial forces in linear members and the resulting membrane stresses are generally much smaller than the bending stresses. The membrane forces are shown in Figure 6.31*b*.

The relative magnitude of the bending and membrane stresses in shells depends on the curvature of the shell. If a shell curves in two directions

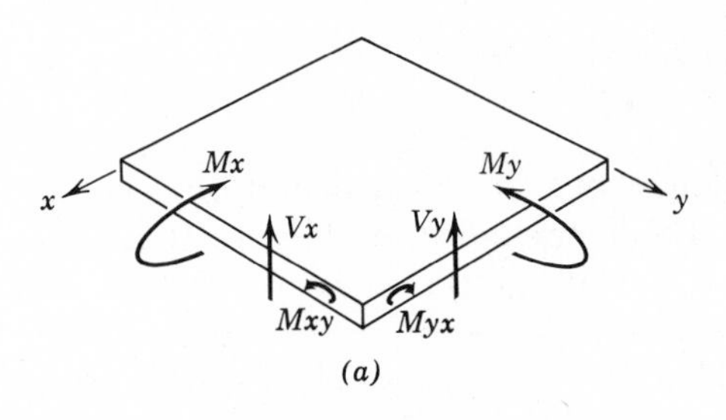


**Fig. 6.31**
Forces in a shell element. (*a*) Bending effects. (*b*) Membrane forces.

(e.g., a spherical shell), the bending effect is very small; such membrane shells therefore have great load-carrying capacity.

The calculation of membrane stresses is relatively simple in common types of shells. If bending occurs as a result of local loads, unfavorable boundary conditions, or flatness, we need to use rather complex bending theories.

You can demonstrate the great difference in membrane and bending effects by compressing an egg by hand in the long direction. If the pressure is uniform, it is virtually impossible to break the egg. It is much easier to exert uniform pressure if compression is applied in the long direction, but if you put a coin in your palm or if the pressure is not uniform enough, the egg breaks easily.

Membrane action is the most efficient type of structural behavior. In theory, practically unlimited areas could be covered by domes (compression shells) or by tension membranes. In practice, however, bending effects at the supports, problems of construction, or cost limit the size of shells.

## 6.7 SUMMARY

The analysis of statically determinate structures was covered in this chapter; it represents the heart of structural analysis. The basic approach is built on the various applications of the equations of equilibrium. The equilibrium of three forces and the construction of the corresponding force triangle is the simplest tool of determinate analysis.

Various structures behave in diverse ways in their task of force transmission, but their analysis is based on the same simple ideas: free body diagrams and their equilibrium.

### *Suggested Reading*

Timoshenko, S. P., and Young, D. H., [1965]: *Theory of Structures,* McGraw-Hill, New York.

Norris, C. H., and Wilbur, J. B. [1960]: *Elementary Structural Analysis,* McGraw-Hill, New York.

Laursen, H. I. [1969]: *Structural Analysis,* McGraw-Hill, New York.

## PROBLEMS

6.1 Draw free body diagrams by making cuts to the right of points *B* and *C* and show the forces acting at these cuts (Figure P6.1).

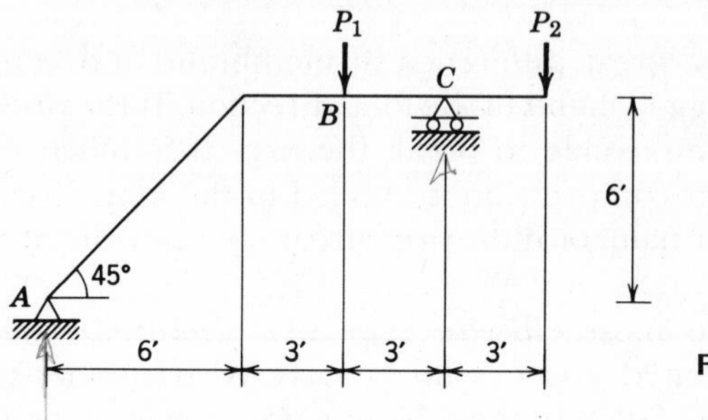


Fig. P6.1

6.2 Calculate the reactions for the structures shown in Figures P6.2*a* to P6.2*i*.

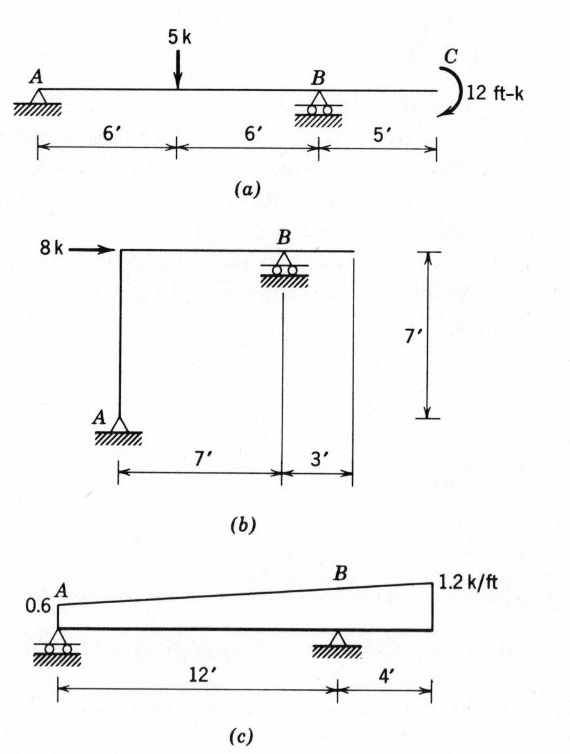

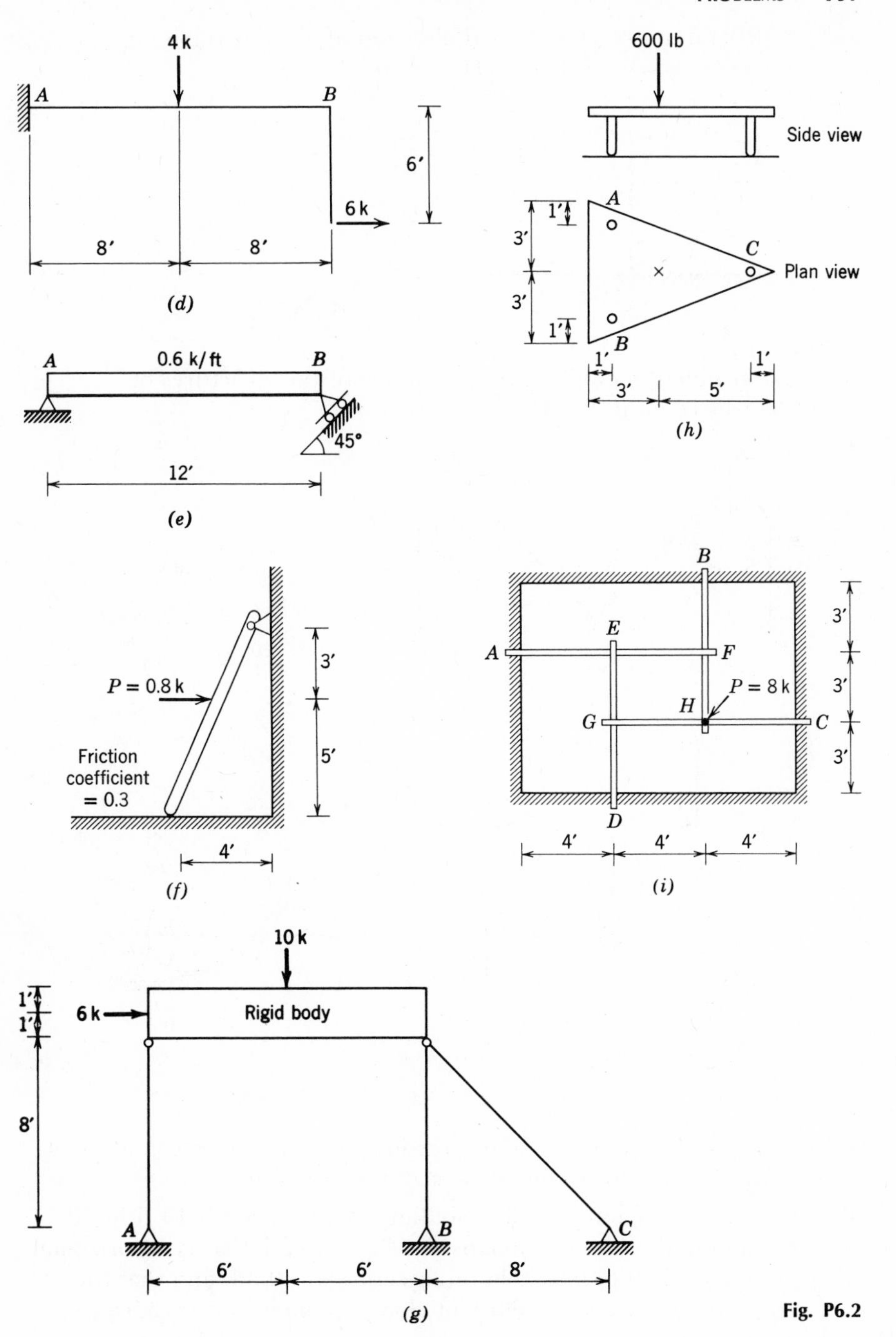
4 k
A
B
6'
6 k
8'
8'
(d)
A
0.6 k/ft
B
45°
12'
(e)
600 lb
Side view
A
1'
3'
C
Plan view
3'
1'
B
1'
1'
3'
5'
(h)
P = 0.8 k
3'
5'
Friction
coefficient
= 0.3
4'
(f)
B
E
A
F
P = 8 k
H
G
C
D
3'
3'
3'
4'
4'
4'
(i)
10 k
1'
1'
6 k
Rigid body
8'
A
B
C
6'
6'
8'
(g)

Fig. P6.2

6.3 Calculate the reactions for the section of a water retaining structure shown in Figure P6.3.

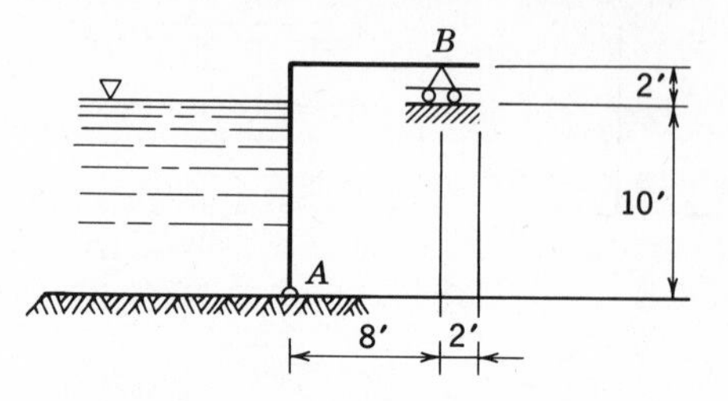


Fig. P6.3

6.4 Determine the equilibrium position(s) of the structures of Figures P6.4*a* to P6.4*d*. Neglect friction.

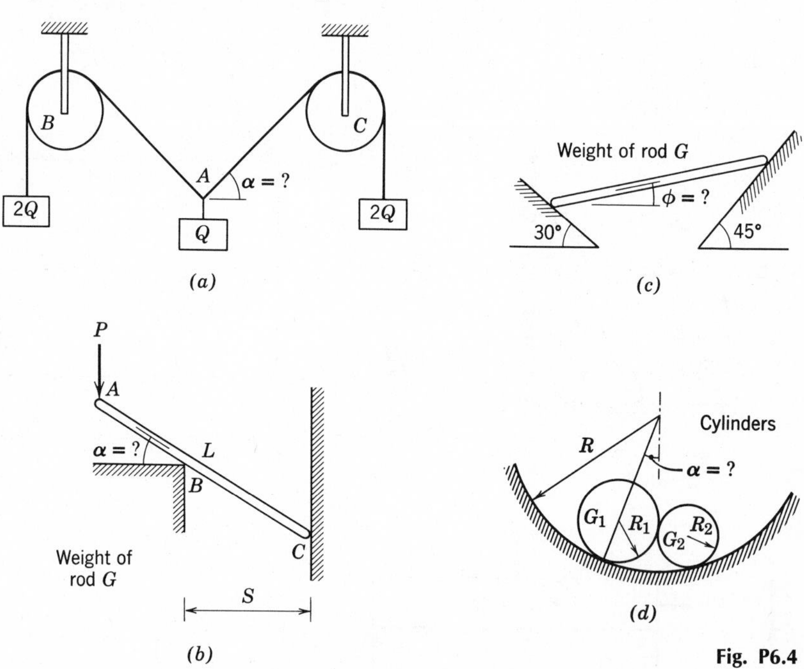


Fig. P6.4

6.5 What is the "hoop" tension force in a circular ring (or in a unit slice of a tube) of radius $R$ caused by internal pressure $p$?

6.6 A concrete retaining wall is holding a bank of earth 10 ft high. Assuming that the soil weighs 100 lb/ft³, that it exerts a horizontal pressure equal to 0.35 that of an equivalent fluid, and that the soil-concrete friction coefficient along the line $AB$ is 0.75, can the wall be in equilibrium? Your statics check should include both sliding and overturning about the toe (point $B$) of the wall (Figure P6.6).

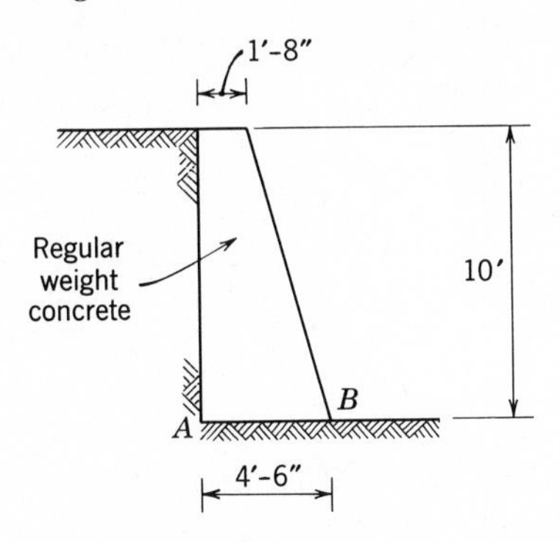


Fig. P6.6

6.7 Calculate the reactions by graphical construction for the following structures:
(*a*) Figure P6.2*b*
(*b*) Figure P6.2*f*
(*c*) Figure P6.4*a* (at $B$ and $C$)
(*d*) Figure P6.4*c*
(*e*) Figure P6.2*g*
(*f*) Figure P6.7

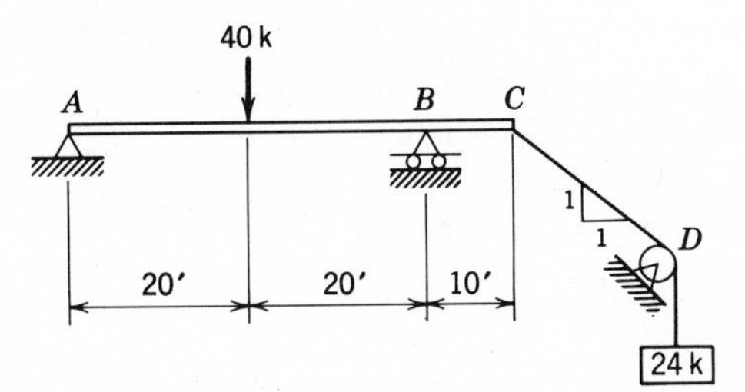


Fig. P6.7

6.8 Calculate the maximum sag in the right span of the continuous cable structure of Figure P6.8. What is the maximum cable tension?

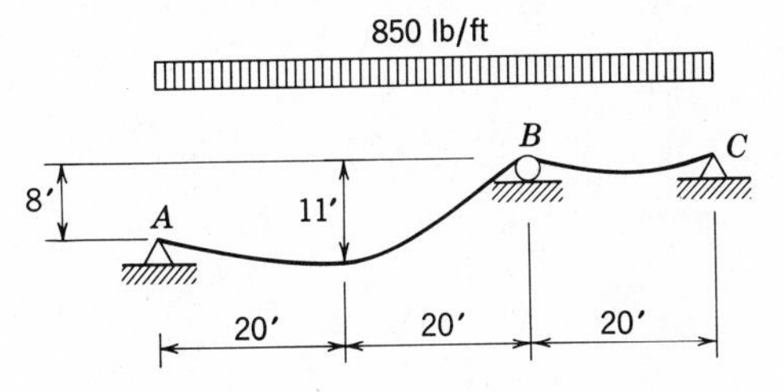


Fig. P6.8

6.9 Calculate the maximum tension in the cable that supports a beam and also carries a uniform load (Figure P6.9).

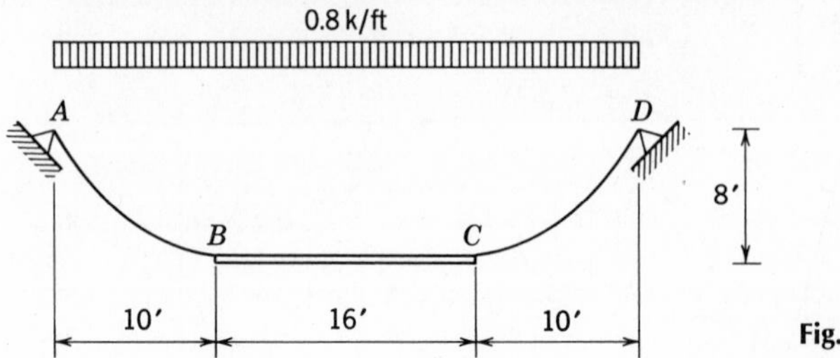


Fig. P6.9

6.10 Compute the ordinates of the cable at the load points. What is the maximum tension? (See Figure P6.10).

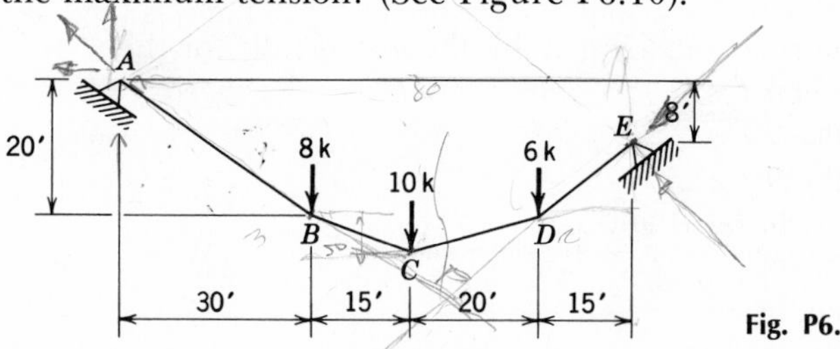


Fig. P6.10

6.11 The cable tower structure shown in Figure P6.11 is in equilibrium under the action of a load of 1 k/ft on the cable and a concentrated load $P$ on the frame. Determine the value of $P$ required for equilibrium. Note that the structure is unstable, but is in neutral equilibrium for this loading.

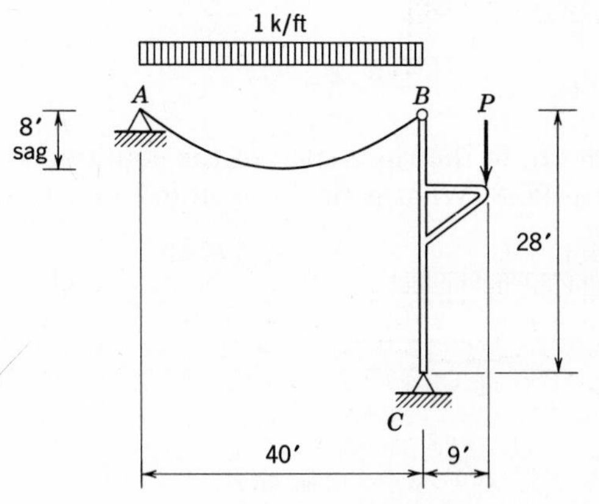


Fig. P6.11

6.12 Calculate the forces in those members of the trusses of Figure P6.12*a* to P6.12*e* that are numbered. All connections are hinges.

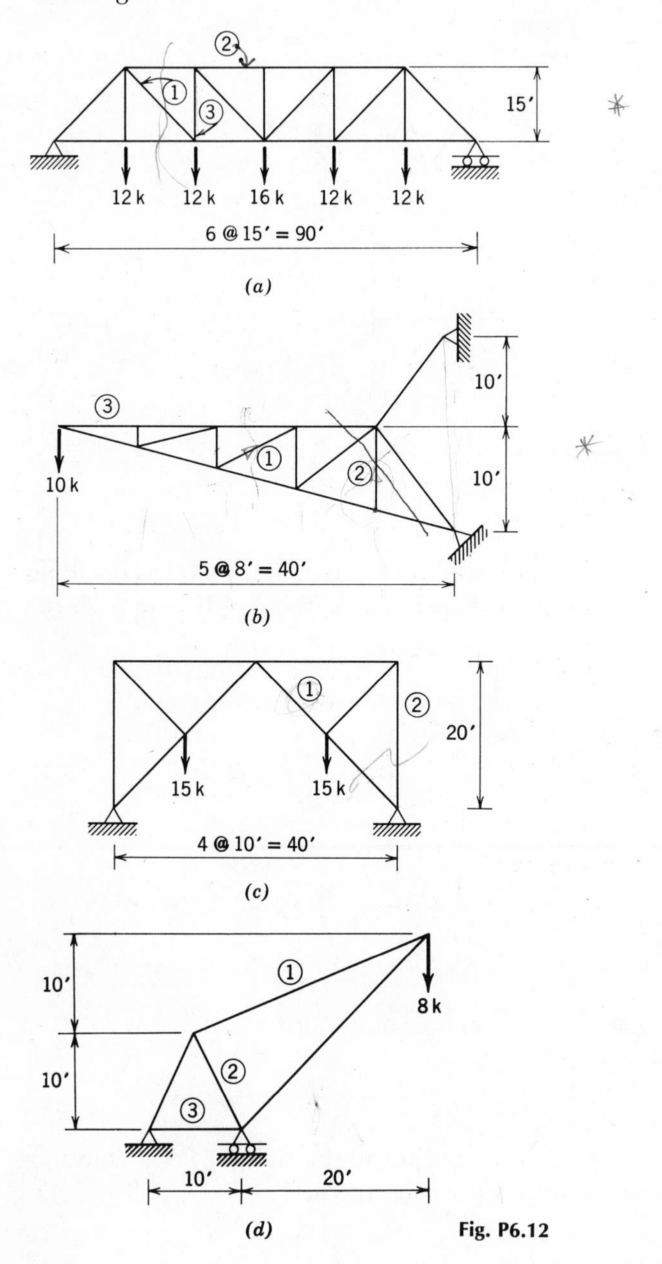


Fig. P6.12

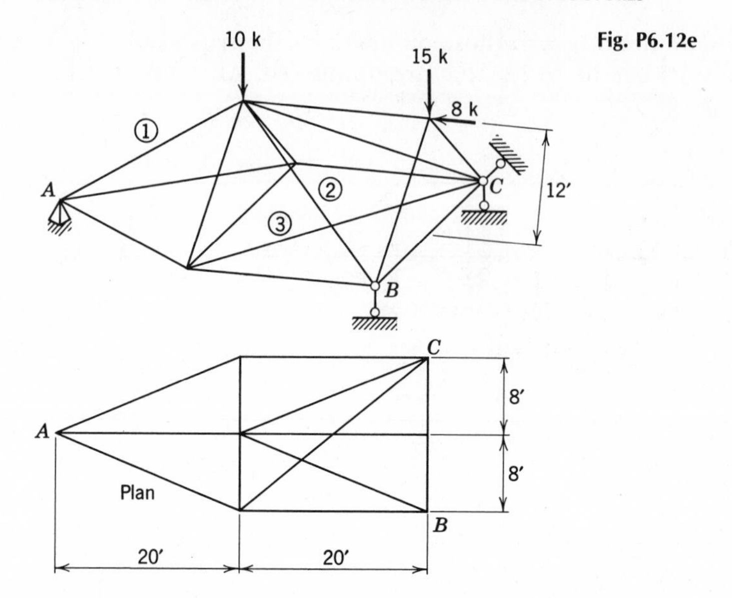


Fig. P6.12e

6.13 Determine the forces in the roof truss (Figure P6.13) resulting from a roof loading of 10 psf. The spacing of trusses is 20 ft.

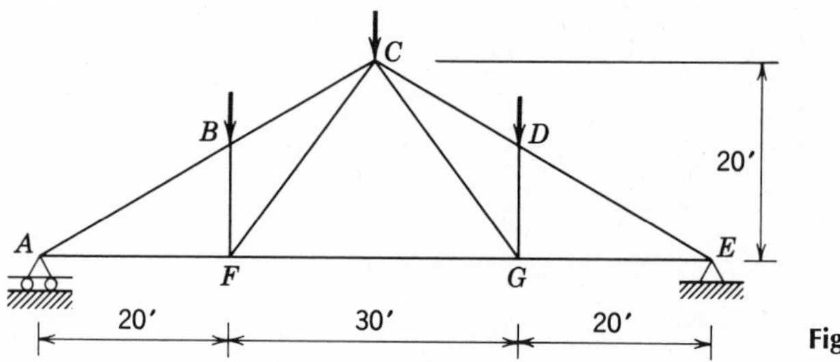


Fig. P6.13

6.14 Solve the following problems graphically:
(*a*) Problem 6.12*d*
(*b*) Problem 6.12*a*
(*c*) Problem 6.13

6.15 Calculate the internal forces just to the right of point $D$ in the three-hinged arch shown in Figure P6.15.

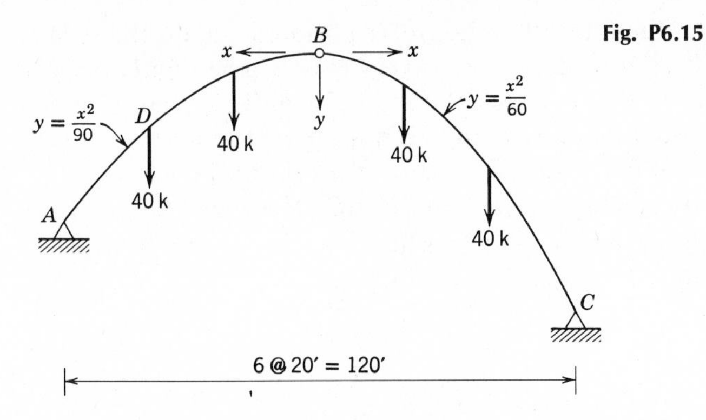


Fig. P6.15

6.16 Calculate the thrust, shear, and moment at point $D$ in the semicircular arch (Figure P6.16) due to the given wind pressure.

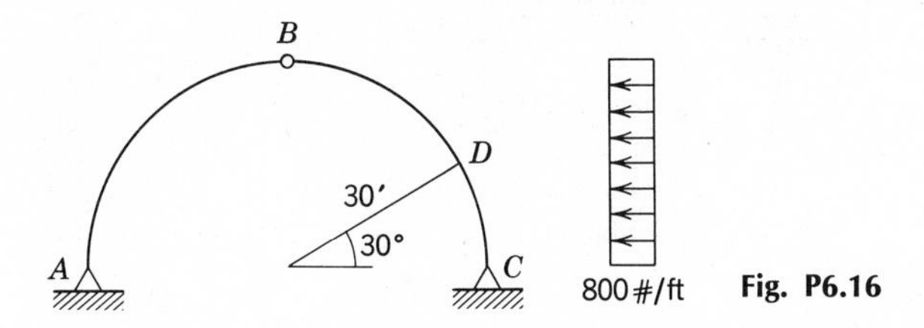


Fig. P6.16

6.17 Calculate the forces in bars 1, 2, 3 of the transmission line tower shown in Figure P6.17. The tower supports three aluminum 3/4-in. diameter cables, which span 400 ft in both directions with

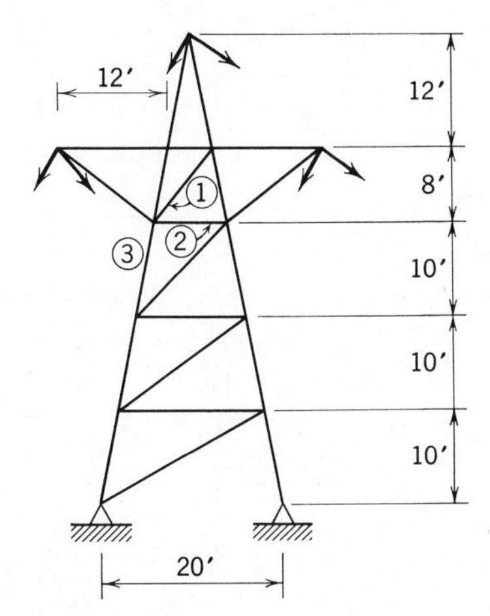


Fig. P6.17

sags of 20 ft. Include the weight of 1 in. radial ice on the cables. (*Hint.* The length of flat cables is very closely given by $L + 8h^2/3L$ and parabolic shape can be assumed instead of the actual catenary shape. Timoshenko and Young [1965] derive the more accurate expressions.) What is the maximum force in the cables?

6.18 An idealized traveling crane is shown in Figure P6.18. Calculate the forces in the major members at a maximum load $W$ which has a safety factor $SF = 3$ against overturning.

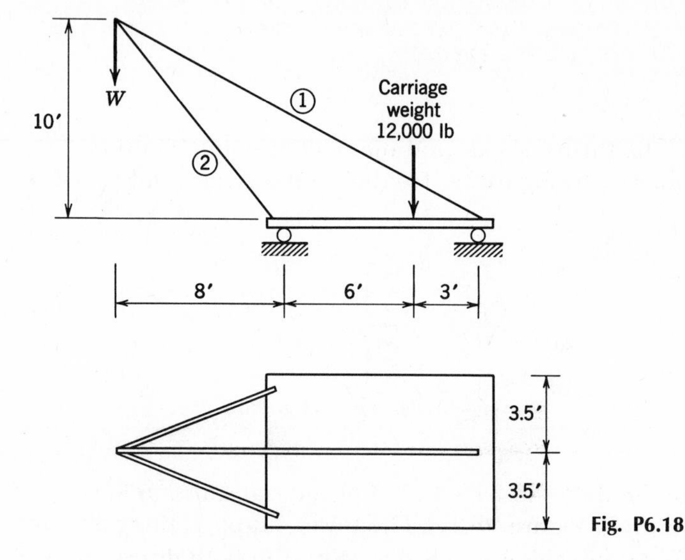


Fig. P6.18

6.19 Calculate the forces in the four bars meeting at joint 7 of the "Pony Truss" of the West Seneca Street Bridge of Ithaca, New York (Figure P6.19). The loading consists of a uniform DL of 40 psf and a 5-ton truck which loads joints 7 and 9 equally. The distance between the two trusses is 21 ft. How does the transverse position of the load affect your results?

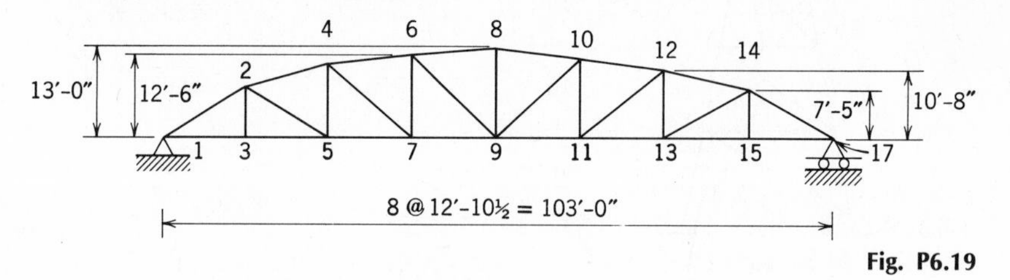


Fig. P6.19

6.20 Solve Problem 6.19 graphically.

6.21 Calculate the internal forces and draw the $V$ and $M$ diagrams for the structures shown in Figures P6.21$a$ to P6.21$e$. Sketch the deflected shape and the $M$ diagram before doing any calculations.

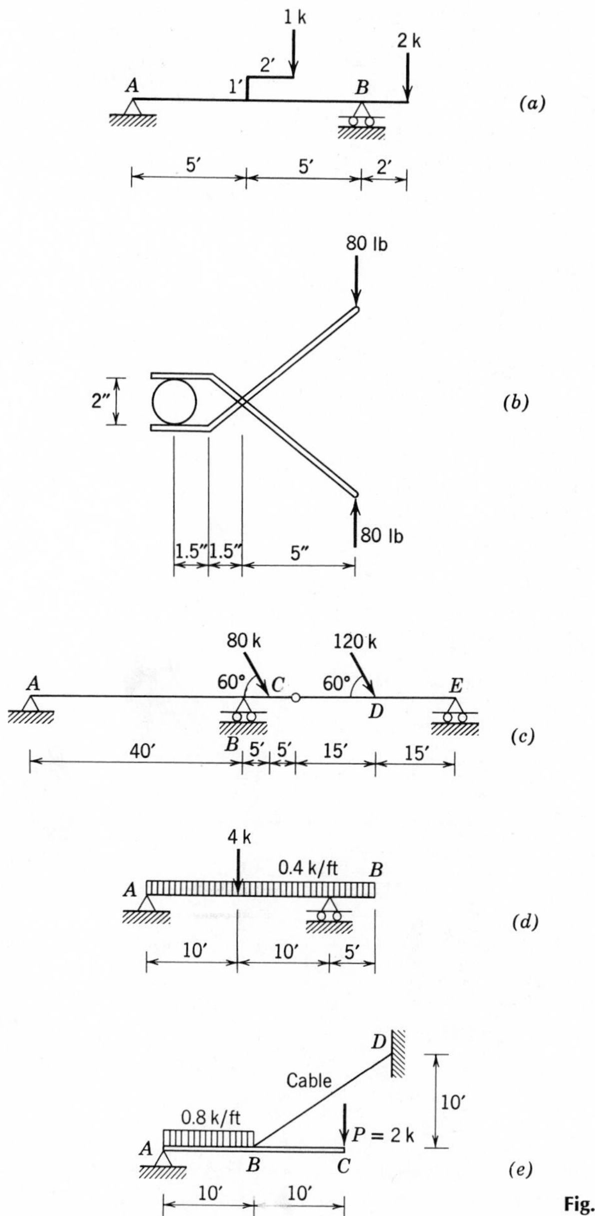


Fig. P6.21

6.22 Calculate the internal forces and draw the $V$ and $M$ diagrams for the structures shown in Figures P6.22*a* to P6.22*d*. Sketch the deflected shape and the $M$ diagram before doing any calculations.

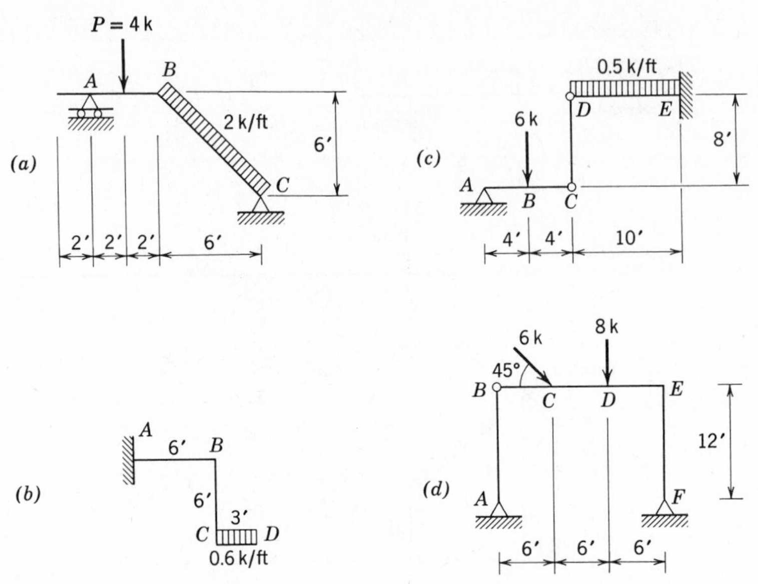


Fig. P6.22

6.23 Sketch the $M$ and $V$ diagrams for the structures shown in Figures P6.23*a* to P6.23*d*.

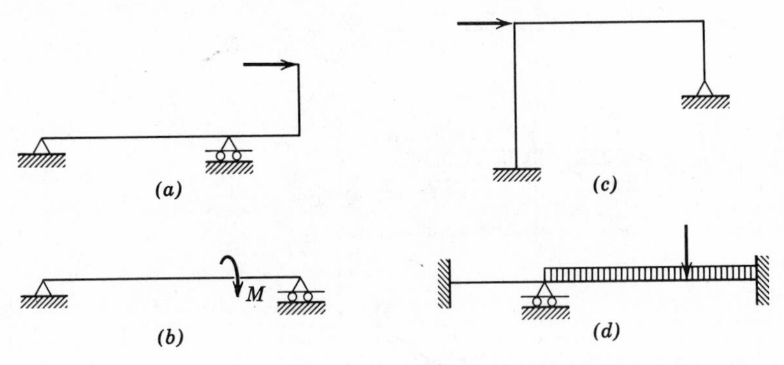


Fig. P6.23

6.24 Sketch the $M$ and $V$ diagrams for the square culvert shown in Figure P6.24. Assume the soil pressure to be uniform.

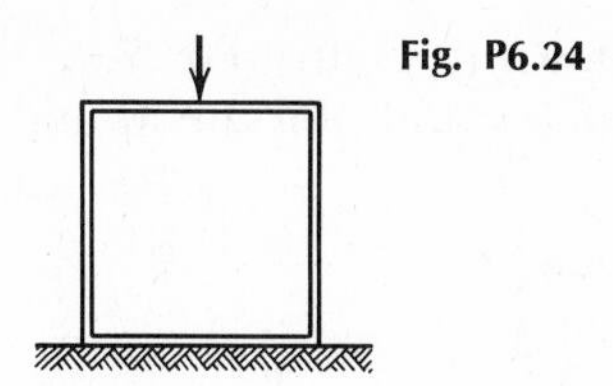
Fig. P6.24

6.25 Analyze the reinforced concrete stairway with landing shown in Figure P6.25. Draw the shear and bending moment diagrams for a 1-ft wide strip of the structure for a combined loading of DL plus a LL of 100 psf.

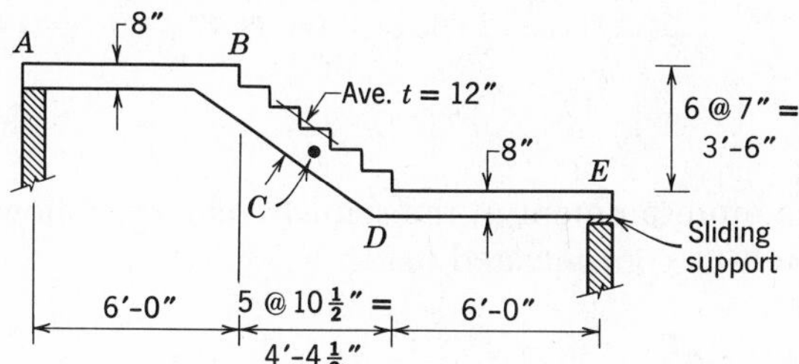

Fig. P6.25

6.26 Draw the $V$ and $M$ diagrams for the water-retaining structures shown in Figures P6.26*a* and P6.26*b*. Assume a 1-ft wide section.

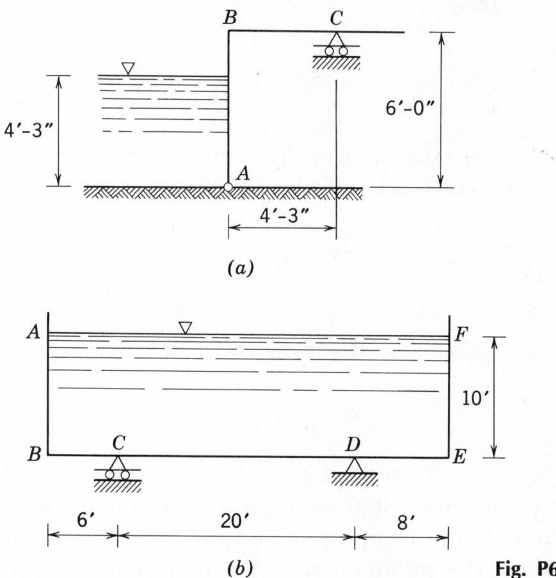

Fig. P6.26

6.27 Analyze the foundation grade beam shown in Figure P6.27. Assume uniform soil pressure. Determine $x$ and plot the $M$ and $V$ diagrams.

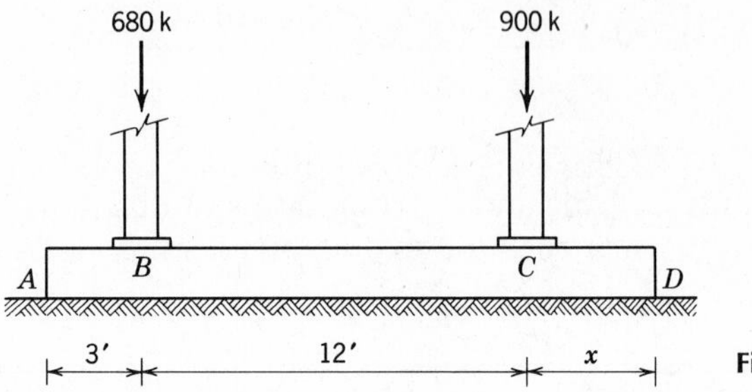


Fig. P6.27

6.28 Determine the maximum moment in the semicircular ring shown in Figure P6.28 due to a concentrated load.

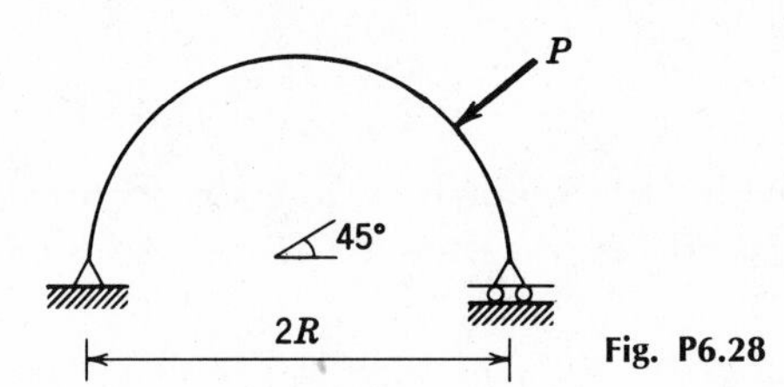


Fig. P6.28

6.29 Determine the maximum moment in the temporary construction support shown in Figure P6.29 and plot the moment diagram.

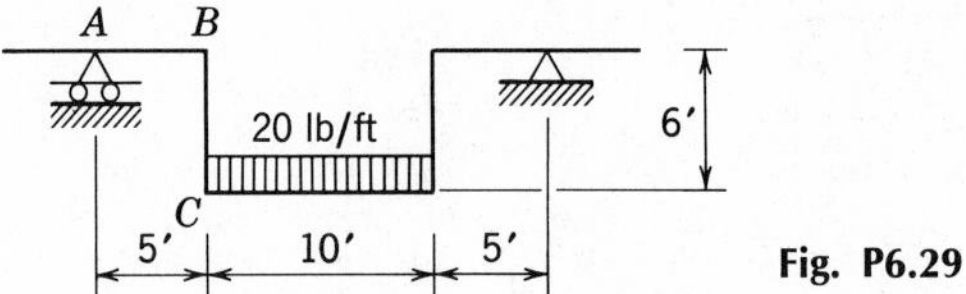


Fig. P6.29

6.30 Analyze the two-story rigid frame (Figure P6.30) under a wind loading of 40 psf. The frame spacing is 20 ft. Use the assumption (to be discussed in detail in Chapter 7) that the inflection points occur at midpoints of the six members. Plot the moment diagram.

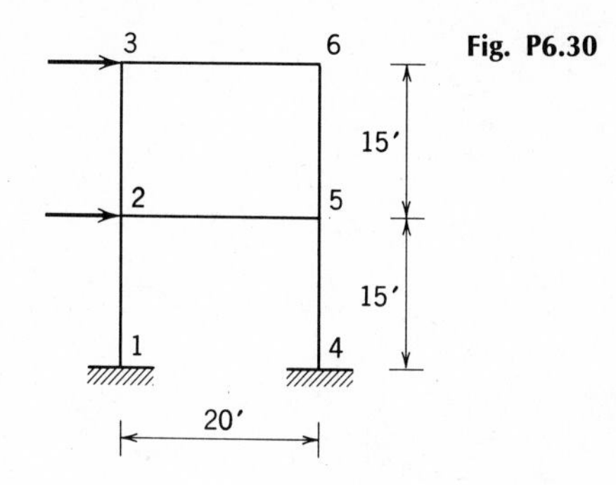

Fig. P6.30

6.31 The shear diagram for a beam subjected to vertical loading (without concentrated moments) is given in Figure P6.31. Determine the loading and the moment diagram.

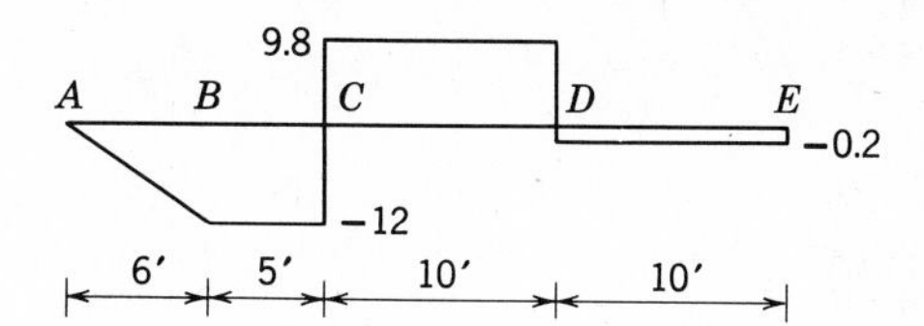

Fig. P6.31

6.32 Using Equations 6.12, calculate the maximum deflection of (*a*) a uniformly loaded simple beam, and (*b*) a cantilevered beam with a concentrated load at its end.

CHAPTER 7

# Approximate Analysis of Statically Indeterminate Structures

Social Sciences Building, Cornell University. *Architect: Skidmore Owings and Merrill*
*Photo by Ezra Stoller © ESTO*

# 7

# Approximate Analysis of Statically Indeterminate Structures

The design and analysis of statically indeterminate structures becomes much more interesting and challenging than that of determinate structures because the analysis depends on a knowledge of member proportions, which are unknown at the time the design-analysis cycle is begun. This fact will be appreciated more fully after you have studied the analysis of statically indeterminate structures using an "exact" method, as covered in Volume 2. The dependence of analysis on member proportions poses quite a dilemma for the engineer: he must know the answers before he can begin! The experienced engineer meets this problem in a number of ways. He can usually judge from experience the approximate member proportions required without performing an analysis. He can estimate dead load in a similar fashion. Most important, he can quickly perform an *approximate analysis* in order to arrive at an estimate of member sizes.

Such an approximate analysis is needed for a number of reasons. We have mentioned its use in estimating member sizes. The designer can make a preliminary cost estimate on the basis of an approximate analysis. He may compare two or more designs on this basis. He can develop a feel for the behavior of his structure. With a determination of member sizes, he can give a good impression of the appearance of the structure to an architect, or perhaps to a citizen's group interested in the appearance of a proposed bridge. In short, approximate analysis is a basic planning tool for decision making on form, materials, cost, and structural proportions. It is also an excellent way of checking the more elaborate computations involved in an exact analysis. We should point out here that no method is really exact, but we often refer to an elastic linear analysis that satisfies compatibility and equilibrium as an "exact" method.

This chapter is concerned with the approximate analysis of a number of structural types that are statically indeterminate and are assumed to behave elastically. We discuss approximate analysis before exact analysis of indeterminate structures because we believe that in this way you can become familiar with a broad range of structures and their behavior, and at the same time gain further insight into equilibrium analysis.

It is important to note at this stage that the elastic linear analysis of indeterminate structures involves member deformations and therefore depends on member stiffnesses. Most approximate methods only roughly account for stiffness effects. When members share a load in an indeterminate manner, the stiffer members carry a greater share. We suggest strongly that you read about the effect of stiffness and of the behavior of indeterminate structures in the article by Cross [1936].

## 7.1 EQUILIBRIUM: THE KEY TO APPROXIMATE ANALYSIS

Every successful structure must be capable of reaching a state of stable equilibrium under its applied loads. This requirement is true regardless of whether the structure is elastic or inelastic, linear or nonlinear, statically determinate or indeterminate. Any method of approximate analysis must include assurance of equilibrium.

In addition to being in equilibrium, most structures reach a state of compatibility of deformation while all parts of the structure are within the range of elastic behavior. Compatibility means that the deformations of the various parts of the structure are consistent with each other (the deformed structure fits together). Elastic behavior is desirable because it is usually accompanied by small displacements, and because the structure returns to its initial position upon unloading. However, compatibility of deformation within the elastic range is not an absolute requirement. A structure may be designed recognizing that equilibrium is reached after some inelastic strain has occurred; such a structure may be entirely satisfactory. Inelastic behavior is discussed in Volume 2, Chapter 17.

The study of approximate methods is best performed by a series of examples, since these methods are often specifically related to a particular type of structure. As we study the examples, you will note a common thread to all: based on a visualization of the way the structure deforms, a number of assumptions are made about internal forces or conditions. The number of assumptions is at least equal to the degree of static indeterminacy. The assumptions and the equations of static equilibrium are then used to analyze the structure. The resulting solution is examined to ensure that it is consistent with the original assumptions and with the understanding of behavior gained during the analysis. The most fundamental fact about all the examples is that equilibrium is maintained. No assumptions inconsistent with equilibrium are admissible.

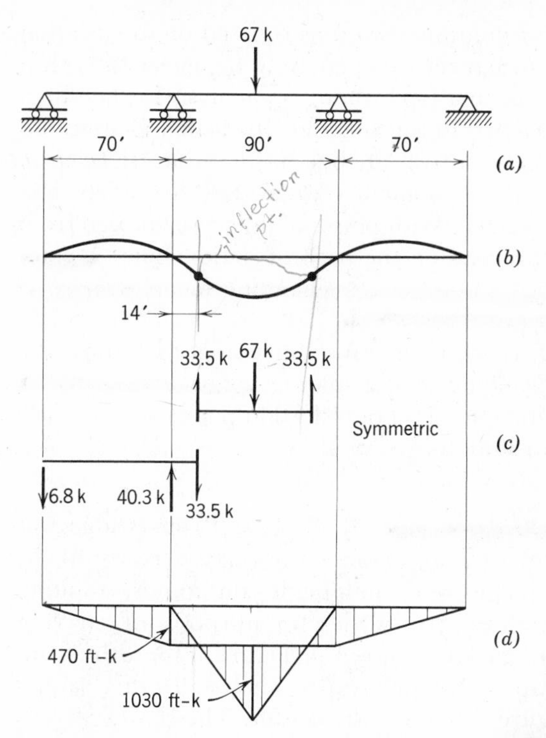


**Fig. 7.1**
Solleks River Bridge girder. (*a*) Layout. (*b*) Deflected shape sketch. (*c*) Statical solution. (*d*) Bending moment diagram.

*ABAM Engineers, Inc.*

## 7.2 ANALYSIS OF CONTINUOUS BEAMS BY SKETCHING THE ELASTIC CURVE

Continuous beams and girders occur commonly in floor systems of buildings and bridges. Because of their frequent occurrence, their approximate analysis is of great interest to us. The three-span bridge girder of the completed Solleks River Bridge, shown in Figure 7.1, is a typical example of this structural type.

The approximate analysis of continuous beams is based on the fact that the elastic curve (deflected structure) can generally be sketched with a fair degree of accuracy without first performing an analysis for bending. When the elastic curve is sketched in this manner, the actual magnitudes of deflection are not accurately portrayed, but the location of *points of inflection* (where the curvature and moment change sign) are easily estimated even on a fairly rough sketch. With points of inflection located from the sketch, the analysis can proceed on the basis of statics alone. A point of inflection, corresponding to a point of zero bending moment, may be thought of as a hinge for purposes of analysis.

There will always be at least enough inflection points to reduce the structure to a statically determinate one, assuming hinges are located at the points of inflection. From the total set of inflection points we select the needed number to achieve a solution by statics.

***Example 7.1 Solleks River Bridge Girder*** The Solleks River Bridge was constructed in a statically determinate manner until it was ready for its concrete deck. The internal hinges were then made continuous, resulting in a statically indeterminate three-span girder for purposes of carrying the live load of vehicles. The girder is shown in Figure 7.1*a*, with a concentrated load located at midspan. An approximate analysis for this loading is desired. The girder of Figure 7.1*a* is drawn to scale. The elastic curve is then sketched in Figure 7.1*b*. Note that it is not necessary to draw a neat, single line; rather a freehand sketching process takes place. Compare the sketched elastic curve with the photograph of a balsa wood model with the same span proportions and loaded in the same manner (Figure 7.2). The sketched shape satisfies the boundary conditions:

1. The ends are simply supported; therefore, the girder can deflect to a position with an end slope as shown.
2. The elevation of the elastic curve is controlled at the four support points.

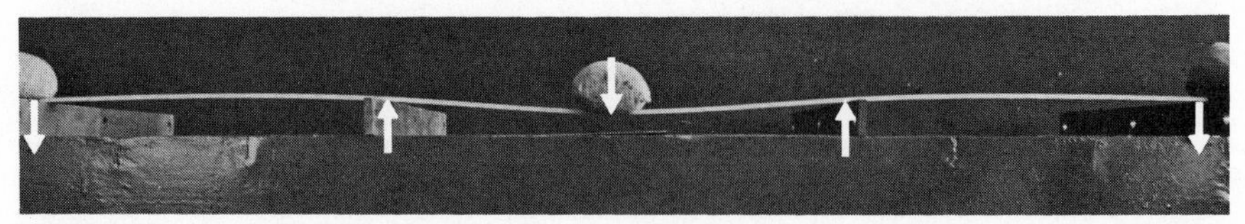

**Fig. 7.2**
Model of Solleks girder.

3. Rotation takes place at each support, but rotation at interior supports is restrained by the outer span members. No sudden change in slope occurs at interior supports because the girder is continuous.

After the sketch "looks right," the inflection points are indicated as large dots on the elastic curve. These are shown in the sketch. They are located by examining the curve and finding the point where it changes from concave to convex.

Figure 7.1*c* indicates free body diagrams of girder segments found by cutting the structure at the inflection points, with the resulting forces computed from equilibrium conditions. Figure 7.1*d* indicates the bending moment diagram that results. Note that the moment diagram is constructed by plotting the simple beam moment diagrams for each span first, and then superimposing the diagram of support moments.

Sketching the elastic curve is an art, and can be learned only by practice and experience. This ability is greatly improved if some experimentation is carried out with the aid of simple wire, paper, plastic, or balsa wood models. We suggest that you attempt to gain as much experience as possible in sketching elastic curves for the example structures in this chapter and for structures appearing in other parts of this book. Benjamin [1959] provides more detailed coverage for those who wish to pursue this topic further.

Before attempting more examples, it is worthwhile to examine a few simple facts of behavior that provide a guide to the deflected structures approach, as it is called by Benjamin. The location of inflection points in any given span can be studied profitably. As an example, consider the fixed-end beam of Figure 7.3*a*. Its elastic curve (from an exact analysis) is shown in Figure 7.3*b* for the uniform load. An exact analysis results in the bending moment diagram shown in Figure 7.3*c*. The inflection points are at a distance $0.21l$ from each end. If we relax the fixity at the right end, both inflection points move to the right, that is, in the direction of *reduced stiffness*. The right inflection point moves all the way to the end as the fixity is reduced to zero as in Figure 7.3*d*, while the left inflection point moves to the left quarter point. *Inflection points move toward places of reduced stiffness, and movement is within narrowly defined limits.*

Now consider the unloaded span of Figure 7.4*a*, with end moments $M_1$ and $M_2$. If $M_1$ and $M_2$ are of opposite directions as in the figure, there are no inflection points in the span. If they act in the same direction as in Figure 7.4*c* (e.g., both being counterclockwise), there is one inflection point. There are no other possibilities for an unloaded span. *No more*

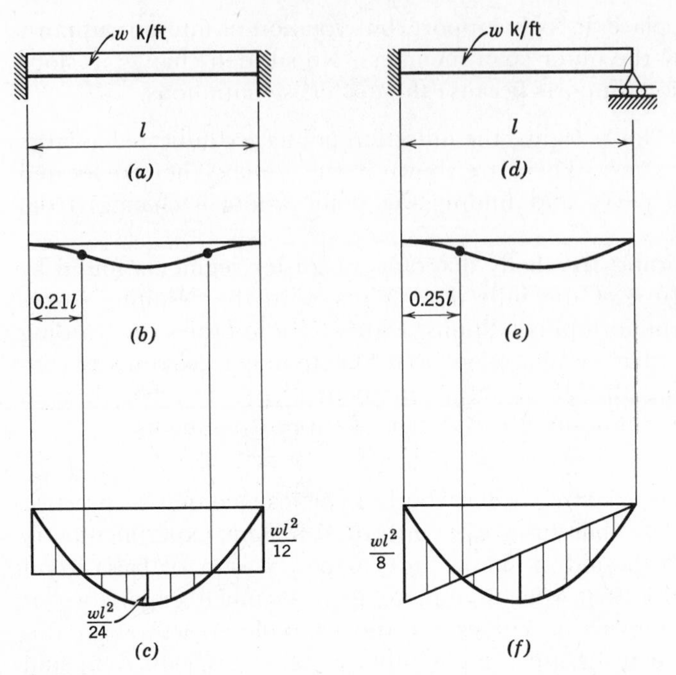


**Fig. 7.3**
Inflection point locations.

*than one inflection point may occur in any unloaded span.* If the span is loaded, as in Figure 7.4*e*, with all loads in the same direction, then up to two inflection points may occur in the span. Note the corresponding bending moment diagrams.

In addition to a study of the behavior of the points of inflection, recall that bending moment is proportional to curvature; from Equation 6.9

$$1/R = M/EI$$

where $R$ is the radius of curvature, $M$ the moment, and $EI$ the flexural stiffness. For members with constant $EI$, maximum moment must occur at points of maximum curvature, zero moments at points of zero curvature, etc.

Knowledge of these simple facts about the number and approximate location of inflection points, in addition to giving you a feel for structural action, should allow you to sketch a reasonable elastic curve for any continuous beam structure. The analysis follows simply and directly once the inflection points are located. The following examples illustrate

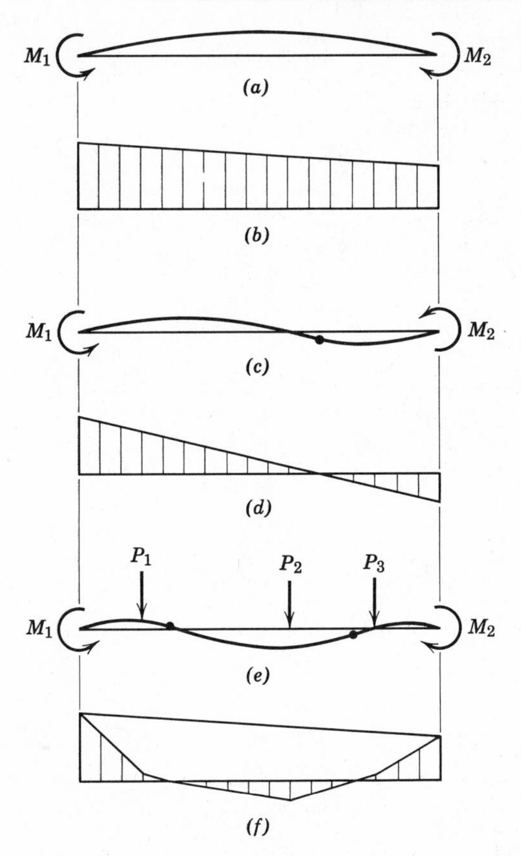


**Fig. 7.4**
Number of inflection points in a span.

the techniques, but real progress can only be made if you attempt to sketch deflected shapes for the examples *before* undertaking a detailed study of the results.

***Example 7.2 Propped Cantilever with Concentrated Load*** Draw the bending moment diagram for the propped cantilever beam shown in Figure 7.5*a*.

The elastic curve is sketched below a scale drawing of the beam. The location of the inflection point is determined by scaling from the sketch, as illustrated in Figure 7.5*b*. Once the inflection point is located, the structure is cut at this location to form the free body diagrams of Figure 7.5*c*. Equilibrium conditions yield the forces indicated on the free body diagram. The moment diagram is constructed as for a statically deter-

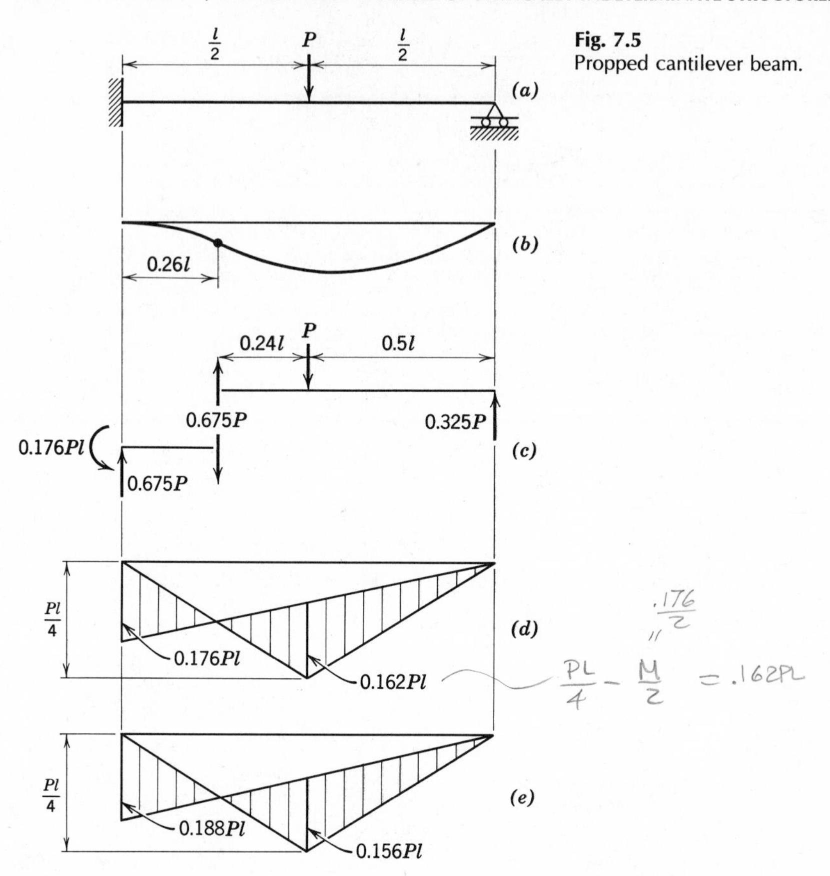


**Fig. 7.5**
Propped cantilever beam.

minate structure, and is shown in Figure 7.5*d*. Figure 7.5*e* indicates the exact bending moment diagram for comparison. The error of 6% at the support is well within desirable limits for an approximate analysis of this type.

***Example 7.3 Continuous Beam with a Concentrated Load*** The bending moment diagram for the beam shown in Figure 7.6*a* is to be drawn. The sketching starts at the cantilever end. We know that the beam rotates clockwise at the right support. Starting here we work back along the beam. Unloaded spans have no more than one inflection point. The only possible smooth curve with a horizontal tangent at the left end is one with a change of curvature in each span. The resulting curve is shown in Figure 7.6*b* with the inflection points located by scaling. The bending moment diagram can be drawn directly as in Figure 7.6*c*, since we know from our study of

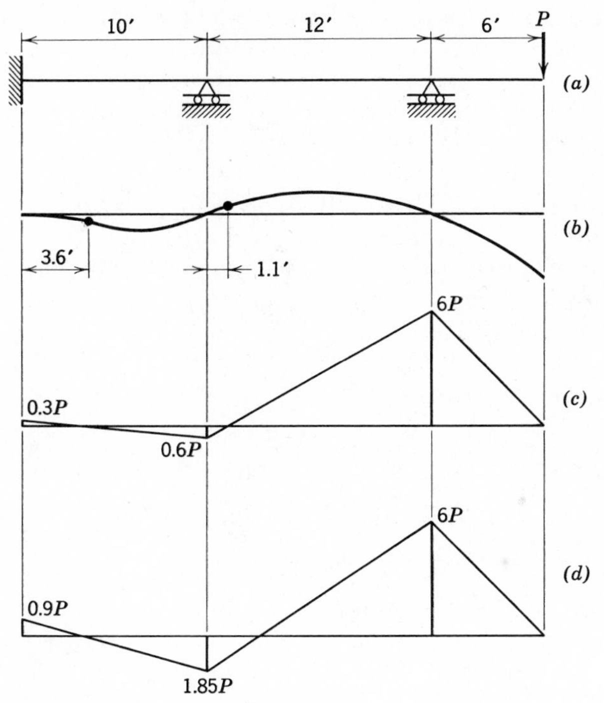


**Fig. 7.6**
Continuous beam: concentrated load.

bending moment diagrams in Chapter 6 that the diagram is linear in an unloaded region. The moment at the right support is determined by statics, with a value $6P$. A straight line drawn through the middle span inflection point (where moment is zero) intersects the support line at a value $0.6P$. The extension in the left span, again through the inflection point, intersects the left support line at a value $0.3P$. We know that curvature is linearly proportional to moment for constant $EI$. The moment diagram and the deflected shape must follow this relationship. If discrepancies are noted, we can make necessary corrections until the elastic curve and the resulting moment diagram appear compatible. Figure 7.6*d* indicates the results of an exact analysis for comparison. Note that the large percentage errors are at regions of small bending moment.

***Example 7.4 Continuous Beam with Uniform Load*** The bending moment diagram is required for the continuous beam of Figure 7.7*a*. Sketching proceeds from the loaded end. The loaded span may have only one inflection point because no moment is developed at the right end. The elastic curve is drawn in Figure 7.7*b*, with inflection points located. If you locate more inflection points in any span than those shown, you will find it impossible to draw a bending moment diagram with straight line segments in the unloaded spans. The inflection point in the loaded span would be at a distance $0.25l$ if the span were fixed at its left support against rotation, as may be recalled from the study of Figure 7.3*d* and 7.3*e*. Since fixity is not the condition, the inflection point must be to the left of this position because it moves toward the region of reduced stiffness.

The moment diagram follows by statics from the locations of the inflection points. It is indicated in Figure 7.7*c*, with the exact moment diagram plotted in Figure 7.7*d* for comparison.

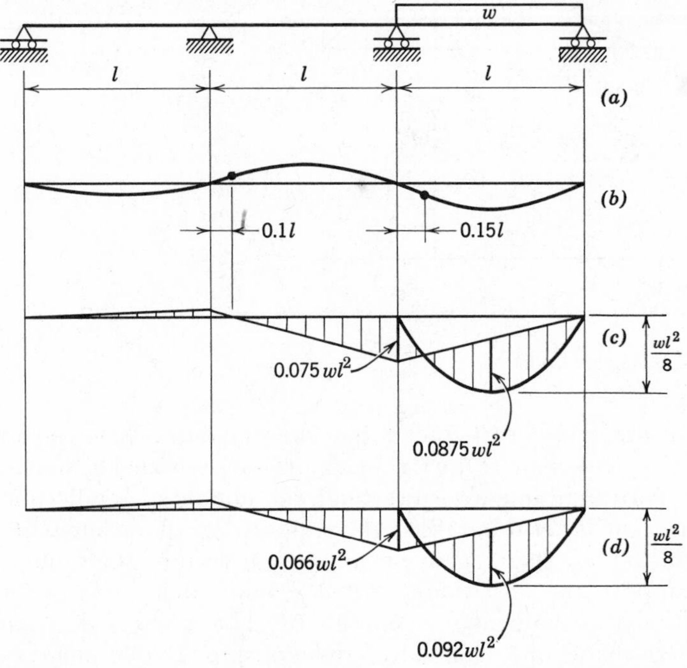


**Fig. 7.7**
Continuous beam: distributed load.

## 7.3 STRUCTURES WITH RECTANGULAR FRAMING AND RIGID JOINTS UNDER LATERAL LOADING

Many structures consist of a number of members connected to form a rectangular rigid framing system. In such a system the joints maintain their original 90 degree angles between framing members. The Vancouver

**Fig. 7.8**
Vancouver International Airport terminal. *Phillips, Barrett, Hillier, Jones, and Partners*

Air Terminal Building is such a structure. Its highly irregular, yet rectangular framing system is shown in the photograph of Figure 7.8. The structure carries gravity loads directly down the columns to the foundations, but lateral wind or earthquake loading, which tends to push the structure sideways, causes bending in both beams and columns. The Cornell University Social Sciences Building has a different form of rectangular framing called a *Vierendeel truss,* shown in Figure 7.14. In this system, the framing is also rectangular but only some of the vertical members extend down to the foundation. Even gravity loads cause bending in both beams and columns. Notice that the Vierendeel truss is not a pin-jointed truss, but a rigid frame.

An understanding of the approximate analysis of rectangular framed structures is best achieved by first considering a doubly symmetric rectangular framed panel of four members rigidly jointed at their ends, as in Figure 7.9*a*. The frame is loaded as shown in the figure and deforms as indicated in Figure 7.9*b*. We may note several facts about the elastic curve shown in the figure.

1. Horizontal and vertical lengths are unchanged, since axial strains are small compared with bending strains.
2. All joints maintain their initial right angles.
3. Since the span between joints is unloaded, it may have at most a single inflection point (this was illustrated in Section 7.2).

The only elastic curve consistent with these observations is the one shown in Figure 7.9*b*. The symmetry of the structure demands that the single

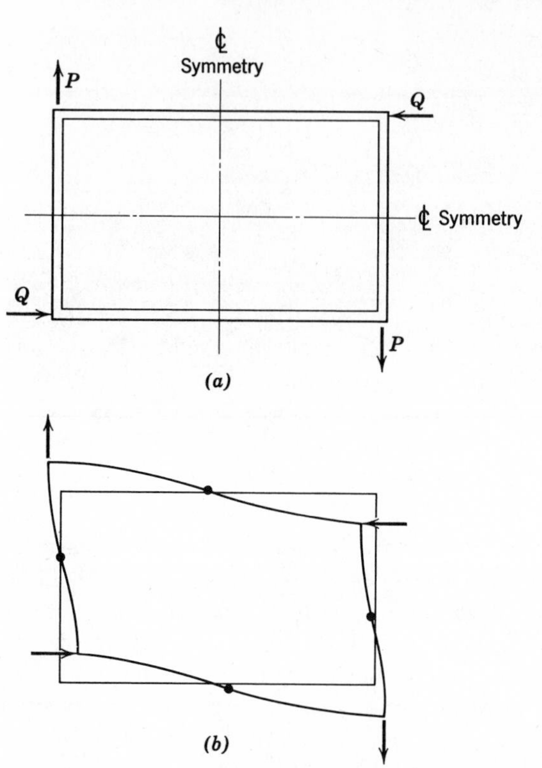


**Fig. 7.9**
Rectangular frame panel.

inflection point be at the midspan of each member, and the antisymmetry of the loading results in an antisymmetric elastic curve.

Actual rectangular frameworks are more complex than the single panel, but when they deform under loads that impose panel shears, we may usually assume with good accuracy that the inflection points occur at midlength of vertical and horizontal members. One exception to this statement occurs in the case of structures with their columns pinned at the base. The structure of Figure 7.13 is such a case. The only inflection point in the column in the lowest story must be at the pinned base, so an additional one at midheight of the first story would be incorrect. The assumption of the location of inflection points reduces the degree of indeterminacy by the number of points assumed.

Additional assumptions are needed in most cases to reduce the analysis problem to one of statics. The nature of these assumptions depends on the structure in question. If the behavior of the structure is dominated entirely by shearing action across the panels, as for a low frame subjected to lateral load, we commonly assume that each panel has equal shear capacity. This means that each vertical member carries shear in proportion to the number of panels of which it is a member (either one or two); for a regular frame like the one shown in Figure 7.10*a*, the exterior columns carry half the shear of the interior columns. The total shear across all columns in a given story must, by equilibrium, equal the total applied shear to that story. The assumptions of inflection point locations and shear distributions, taken together, constitute the *portal method* of analysis, and are valid for frames such as that in Figure 7.10*a*.

Although panel shearing action is significant, the behavior of tall structures is affected by the overall bending of the structure as a beam cantilevered up from the ground, as in Figure 7.10*b*. The figure indicates this overall bending effect but we recognize that panel shearing action also takes place. As a result of the overall bending caused by lateral loads, the columns act as the fibers in a beam, with tension on the windward side and compression on the leeward side, in proportion to their distance from the centroid of the column group. Column axial forces can be computed from an understanding of this behavior. The assumptions of inflection point locations and column axial forces, taken together, constitute the *cantilever method* of approximate frame analysis.

The portal and cantilever methods of analysis are discussed by Benjamin [1959] and by Norris and Wilbur [1960]. In each of these methods, the number of assumptions regarding shears or column axial forces and the number of assumed inflection points together yield enough conditions so that the analysis can proceed on the basis of statics.

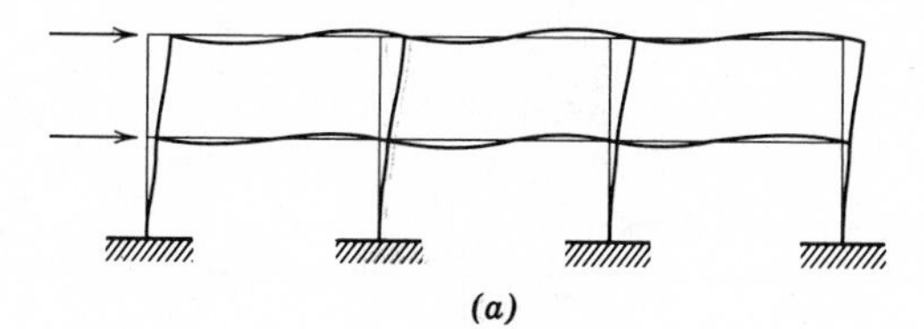


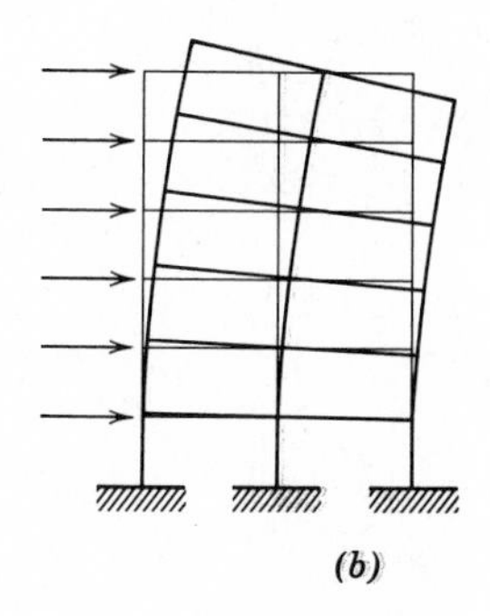


**Fig. 7.10**
Rectangular frames.

***Example 7.5 Approximate Analysis of a Low Building Frame by the Portal Method*** Structures with heights less than their width may be termed "low," and for these types, the method of analysis based on dominance of panel shear behavior under horizontal loads is appropriate. The building of Figure 7.11*a* is an example of such a structure, laterally loaded with wind forces. A complete analysis for the structure is desired (although in many cases only the bending moments might be needed).

The first step in the analysis is to place the assumed inflection points at the midpoint of each member. With the assumption of these "hinge locations" the degree of static indeterminacy is reduced from 12 to 2.

The next assumption is with regard to the distribution of shears in each story. The total shear of 20 k in the upper story is shared equally by each panel. Therefore, the left upper column carries its share of the panel shear, 5 k. The center upper column carries a share for each panel, a total of 10 k. The right upper column carries 5 k. The lower panels are treated in the same manner. In this case, the total shear to be distributed is 30 k. We assign 7.5 k to each exterior lower column and 15 k to the interior lower column. Figure 7.11*b* indicates the structure cut into nine free body

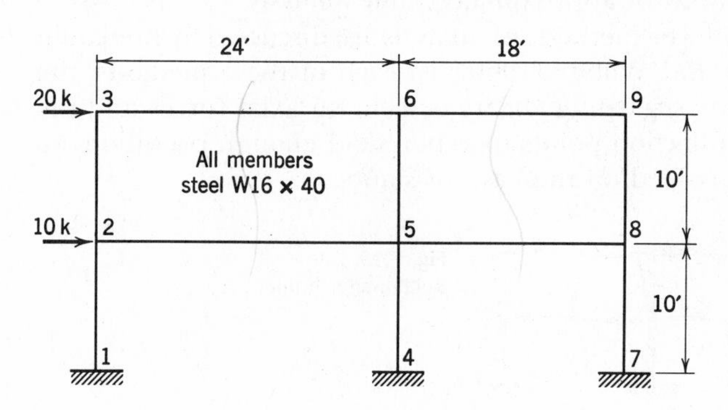


**Fig. 7.11**
Portal method of analysis.

diagrams, with column shears shown at the column inflection points. Note that the structure is cut at each of the inflection points, creating a set of free bodies without internal hinges.

The figure also indicates the unknown beam and column axial and shear forces $P_1$ to $P_{14}$ and the base reactions. All of these unknown forces are determined readily by considering static equilibrium of each free body in turn, starting with one of the upper outside free body diagrams with only three unknowns. The results of such computations are indicated in Table 7.1 with the results of an exact analysis for comparison. The moment diagrams for both the approximate and exact analyses are plotted in Figure 7.12*a* and 7.12*b*. Note that Figure 7.12*b* also indicates the actual inflection point locations (points of zero moment).

**Table 7.1 Internal Actions and Reactions: Portal Frame of Figure 7.11**

| | Approximate | Exact |
|---|---|---|
| $P_1$ | − 2.08 | − 2.21 |
| $P_2$ | 15.0 | 15.44 |
| $P_3$ | 2.08 | 3.41 |
| $P_4$ | − 7.28 | − 5.62 |
| $P_5$ | 7.5 | 5.57 |
| $P_6$ | 5.2 | 3.41 |
| $P_7$ | − 0.70 | − 1.33 |
| $P_8$ | 5.0 | 5.65 |
| $P_9$ | 2.78 | 3.55 |
| $P_{10}$ | − 2.45 | − 3.56 |
| $P_{11}$ | 2.5 | 3.74 |
| $P_{12}$ | 6.95 | 5.64 |
| $P_{13}$ | 2.78 | 3.55 |
| $P_{14}$ | 9.73 | 9.18 |
| $H_{1,2}$ | 7.5 | 9.00 |
| $H_{4,5}$ | 15.0 | 11.62 |
| $H_{7,8}$ | 7.5 | 9.39 |
| $V_{1,2}$ | − 7.28 | − 5.62 |
| $V_{4,5}$ | − 2.45 | − 3.56 |
| $V_{7,8}$ | 9.73 | 9.18 |
| $M_{1,2}$ | 37.5 | 63.95 |
| $M_{4,5}$ | 75.0 | 71.85 |
| $M_{7,8}$ | 37.5 | 64.03 |

It is important to realize that in many analysis problems all of the internal forces are not required. For example, the column bending moment diagram is readily constructed from the column shears since the points of inflection are located.

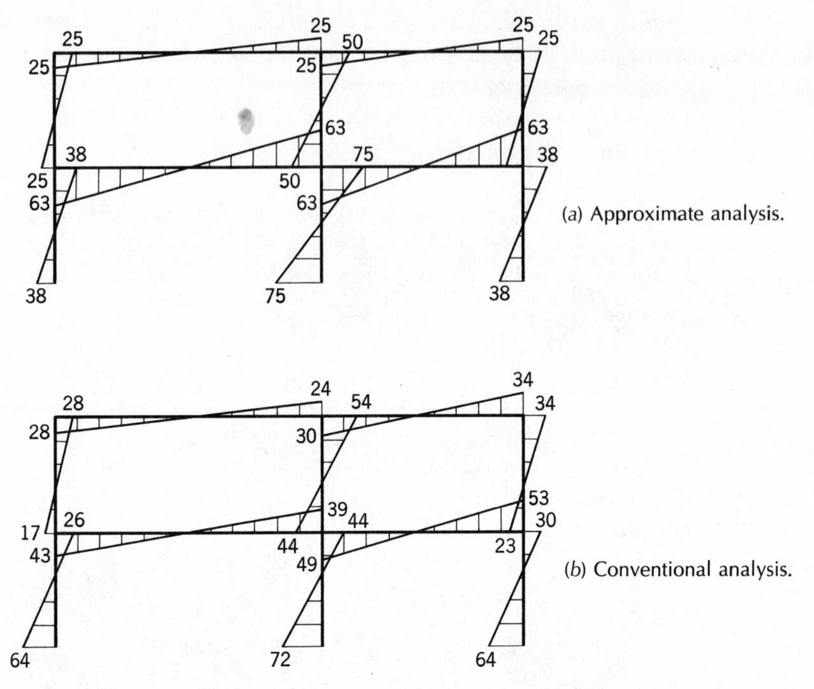


**Fig. 7.12**
Moment diagram: portal frame (ft-k).

***Example 7.6 Approximate Analysis of a Tall Building Frame by the Cantilever Method*** Although the division between classifications "low" and "tall" is rather arbitrary, a rectangular framework structure with greater height than width might be classed as "tall." For structures of this type, a method of analysis based on the assumption that the building is considered to act like a cantilever beam, cantilevered from the ground, is appropriate, although the portal method may also be used. In the cantilever method the column lines are assumed to act as the fibers in a beam.

Consider the tall building frame loaded with wind as indicated in Figure 7.13*a*. Assume that all column areas are equal (you may find it interesting to adapt the method to the case where column areas are unequal, keeping in mind that column strains are assumed to vary linearly with the distance from the centroid of the group of column areas). The analysis begins with the placement of inflection points at midlength of all members. In this case, the columns of the first story are exceptions. They are hinged at the base; therefore, the inflection points are moved down to the base.

For such a tall building, column strains resulting from overall bending action are assumed to affect behavior. We assume that the building is a laterally loaded cantilever beam with the cross section as indicated in Figure 7.13*b*. The centroid of the column group (and the axis of bending) is located in the figure. Overall bending with a linear strain distribution results in column forces varying linearly from the centroid. Let column 11 have tension $X$. Then column 12 has compression $(2.67/17.33)X = 0.154X$. Column 13 has compression $(14.67/17.33)X = 0.845X$. The total moment acting on a section through the first story inflection points (the base in this case) is resisted by these column forces. Therefore, equilibrium yields

$$12(50) + 10(40) + 10(30) + 10(20) + 5(10) = X(17.33) + 0.154X(2.67) + 0.845X(14.67)$$

$$1550 = 30.1X$$

$$X = 51.5 \text{ k}$$

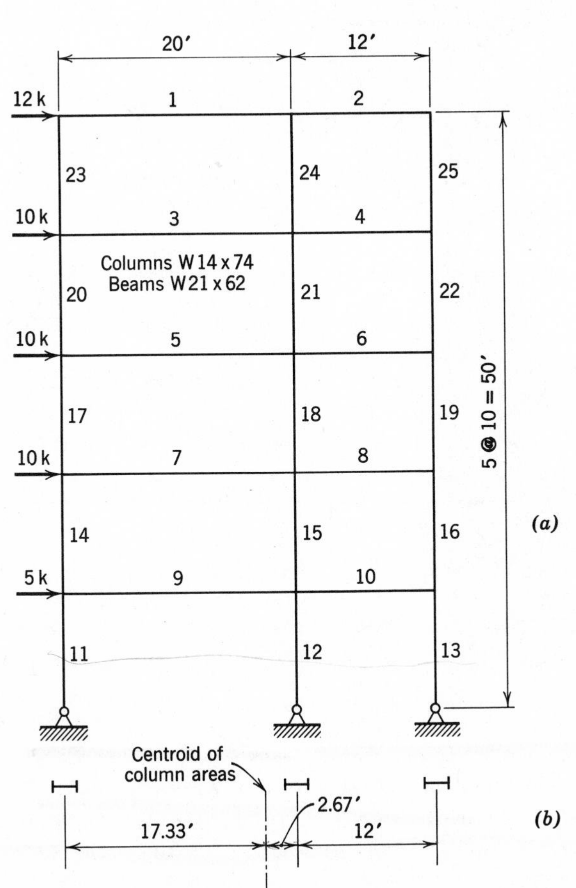


**Fig. 7.13**
Tall building frame.

The other axial forces at the base are then readily computed, and are indicated in Table 7.2. Column axial forces may be similarly computed at each inflection point level by finding the moment of lateral forces about the level in question, and by assuming columns to act as beam fibers. All forces are given in Table 7.2.

**Table 7.2 Internal Forces: Tall Building of Figure 7.13**

| | Member | Axial Force (k) (tension +) | | Shear Force (k) (clockwise on member +) | | Bending Moment (ft-k) (clockwise on end of member +) | | |
|---|---|---|---|---|---|---|---|---|
| | | Approximate | Exact | Approximate | Exact | Approximate | Exact | |
| | | | | | | | Left or Bottom | Right or Top |
| Beams | 1 | −8.0 | −8.3 | −2.0 | −2.3 | 20 | 23.7 | 22.6 |
| | 2 | −2.1 | −2.6 | −1.7 | −2.1 | 10 | 9.2 | 16.3 |
| | 3 | −6.6 | −8.0 | −5.7 | −4.5 | 57 | 46.7 | 42.2 |
| | 4 | −1.2 | −3.5 | −4.8 | −6.8 | 29 | 36.9 | 44.3 |
| | 5 | −7.0 | −7.8 | −8.9 | −6.4 | 89 | 68.0 | 59.8 |
| | 6 | −2.8 | −3.3 | −7.5 | −11.9 | 45 | 65.7 | 77.3 |
| | 7 | −7.2 | −10.0 | −11.8 | −8.6 | 118 | 90.9 | 80.8 |
| | 8 | −2.7 | −2.8 | −10.0 | −18.2 | 60 | 102.7 | 115.3 |
| | 9 | −1.7 | 0.2 | −23.1 | −13.9 | 231 | 154.2 | 123.5 |
| | 10 | 3.1 | −2.3 | −19.5 | −30.8 | 117 | 168.7 | 201.1 |
| Columns | 11 | 51.5 | 35.6 | 16.5 | 13.1 | −165 | 0 | −131.5 |
| | 12 | −8.0 | 34.2 | 24.8 | 18.7 | −248 | 0 | −187.0 |
| | 13 | −43.5 | −69.8 | 5.7 | 15.2 | −57 | 0 | −151.6 |
| | 14 | 28.4 | 21.7 | 13.2 | 7.9 | −66 | −22.8 | −56.3 |
| | 15 | −4.4 | 17.2 | 20.0 | 21.9 | −100 | −105.2 | −113.9 |
| | 16 | −24.0 | −39.0 | 8.8 | 12.2 | −44 | −49.5 | −72.4 |
| | 17 | 16.6 | 13.2 | 10.4 | 7.9 | −52 | −34.7 | −44.5 |
| | 18 | −2.6 | 7.7 | 15.5 | 14.7 | −77 | −69.7 | −77.1 |
| | 19 | −14.0 | −20.8 | 6.0 | 9.4 | −30 | −42.9 | −51.1 |
| | 20 | 7.7 | 6.8 | 7.4 | 5.7 | −37 | −23.5 | −33.4 |
| | 21 | −1.2 | 2.1 | 11.3 | 10.2 | −56 | −48.4 | −53.8 |
| | 22 | −6.5 | −8.9 | 3.3 | 6.1 | −16 | −26.2 | −34.7 |
| | 23 | 2.0 | 2.3 | 4.0 | 3.7 | −20 | −13.3 | −23.7 |
| | 24 | −0.3 | 0.2 | 5.9 | 5.7 | −29 | −25.3 | −31.8 |
| | 25 | −1.7 | −2.1 | 2.1 | 2.6 | −10 | −9.6 | −16.3 |

After all axial forces have been determined, the analysis proceeds on the basis of statics, starting at an upper corner and proceeding through each free body diagram in turn. Convenient free bodies are again created by cutting at inflection points, since each cut introduces only two unknown forces.

Table 7.2 contains a summary of the internal forces found in this manner and, for comparison, those found by an exact computer-aided analysis. The approximate case bending moments are easily calculated for any member from the value of shear in that member. The moment at the end is simply the shear (acting at the inflection point) multiplied by the distance from the inflection point to the end. The bending moments for the approximate analysis are indicated in Table 7.2, moment being the same at each end except for the lower story columns where it is zero at the lower end. The bending moments found from the computer-aided analysis are also given. An evaluation of their magnitudes at opposite ends of a member will indicate the location of the inflection point. A careful comparison of these results will aid greatly in visualization of structural action; for example, the location of inflection points and the distribution of column loads.

***Example 7.7 Approximate Analysis of a Vierendeel Truss*** The Vierendeel truss of Figures 7.14 and 7.15 was constructed to provide column-free space under the Cornell University Social Sciences Building. It has a rectangular framing system that must carry vertical loads in bending action of both beams and columns. Only two support points are available to carry the loads to the foundation.

Rather than performing a complete approximate analysis, we will demonstrate here how a minimum of effort can lead to an approximate

**Fig. 7.14**
Cornell University Social Sciences building: Vierendeel truss.

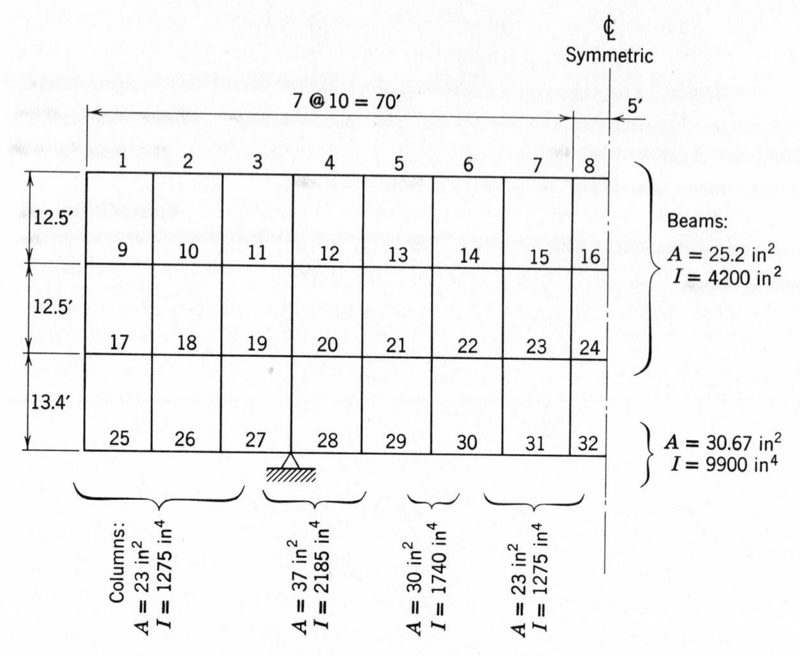


**Fig. 7.15**
Vierendeel truss.

idea of important member forces and moments. We want to find the maximum bending moment in the horizontal members caused by gravity loads. We recognize that the portal method, which involves the allocation of shears, will give us directly the shears in any horizontal member. Knowledge of the shear is sufficient to compute the bending moment. Examination of the structure indicates that the maximum shear occurs just to the right of the support, that is, across members 4, 12, 20, and 28. The shear across these members equals the total vertical load to the right of them. This load consists of 24 k applied to each of 16 joints, giving a total shear of 384 k, which is distributed among the four horizontal members. The portal method would assign twice the shear to interior members 12 and 20 as compared with that assigned to exterior members 4 and 28. This would result in 64 k shear in members 4 and 28 and 128 k in members 12 and 20. In this particular case, however, we can improve on the usual assumptions of the portal method. We note from Figure 7.15 that the lower horizontal member is much stiffer than the others. It will therefore take more shear than that indicated by the portal method. One reasonable assumption would be to assign to the lower member (number 28) the same shear as that carried by interior members. Thus, we assign about 55 k to the upper member (number 4) and 110 k to the others. The inflection points are assumed to lie at midspan. The maximum moment is the shear (110 k) multiplied by the distance from the inflection point to the joint (5 ft), or 550 ft-k. An exact analysis for the same loading, performed with the aid of a computer program gives a value of 345 ft-k in member 4, 503 ft-k in member 12, 602 ft-k in member 20, and 1132 ft-k in member 28. The discrepancy in the latter is caused by the fact that the inflection point was actually located near the right quarter point on member 28. When using approximate methods, one must be aware that they are approximations and that the assumptions may be invalid in some circumstances. In this case, the lack of uniformity of member sizes has contributed to the discrepancy.

## 7.4 ANALYSIS OF SPECIAL STRUCTURES

Although many structures fall into the class of beams or rectangular frames, a discussion of approximate methods of analysis would not be complete without introducing some of the considerations encountered in other structures. This section is not intended to provide detailed instruction in the analysis of the specific examples discussed, but rather to indicate the general approach to approximate analysis, especially the necessity of understanding the basic action of the structure in qualitative terms, and the use of this understanding to make assumptions about conditions (points of inflection) and internal actions and reactions. The importance of equilibrium, and the interpretation of results with care and judgment, cannot be overemphasized.

***Example 7.8 Analysis of a Gable Frame*** The two-hinged gable frame indicated in Figure 7.16 is loaded with a 10-kip concentrated load. We desire an approximate analysis for this structure. We recall from Section 6.4 that three forces in equilibrium must have their lines of action intersect at a point. If there had been an internal hinge at 4 as in a three-hinged frame, the solution would proceed as described in Section 6.4. The reactions $R_1$ and $R_6$ would be drawn in the directions indicated by the dashed line of Figure 7.16*a* intersecting at $O'$. Since no hinge exists in this structure, the member 2-5 is somewhat stiffer than it would be otherwise, and the horizontal components of the reactions are reduced. Consider an extreme case: if the member 2-5 was very stiff, and the columns flexible, 2-5 would act like a simple beam and the horizontal component of the reactions would be small. Reduction of horizontal reactions, while maintaining equilibrium, corresponds to raising the intersection point for the three forces in Figure 7.16*a* as indicated by the broken line. For this indeterminate structure, the point of intersection is not known, but we can guess its location as denoted by point $O$ in the figure. Incidentally, statics will show that the bending moment at any point on the frame is given by the magnitude of the reaction multiplied by the perpendicular distance from its line of action to the frame member (moment at 5 is $R_6$ multiplied by distance $A$-5). Since the structure has a constant cross section, we might anticipate that a desirable design would have approximately equal positive moment at 3 and negative moment at 5. Point $O$ in Figure 7.16*a* is located to approach this condition. The broken line is the line of action of the reactions, and is called the *thrust line* or *funicular line,* since a set of members along this line would have no bending.

With point $O$ located the graphical force polygon is drawn as indicated in Figure 7.16*b*. The known force of 10 k is drawn first, then the reactions are included from their known directions. Their magnitudes are scaled from the figure, with the result that $R_6 = 4$ k and $R_1 = 8$ k. The bending moment at any point is then computed by the method mentioned above; $A$-5 $= 7$ ft, $B$-3 $= 8$ ft, and 2-$C = 3$ ft. The moment at 5 is therefore $4 \times 7 = 28$ ft-k. The moment at 3 is $4 \times 8 = 32$ ft-k, and the moment at 2 is $8 \times 3 = 24$ ft-k. The moment diagram is plotted in Figure 7.16*c*.

For comparison, an exact analysis, based on constant $EI$ for all members, gives $M_5 = 26.8$ ft-k, $M_3 = 36.2$ ft-k, and $M_2 = 21.8$ ft-k. These values agree closely with those found from the approximate analysis. However, one should avoid generating too much confidence in the method based on a single example. This method can give very poor results. The important point is that a rough answer can be found so easily that there is no excuse for estimating the structural sizes with order of magnitude errors.

***Example 7.9 Analysis of a Stiffened Tied Arch*** An interesting structural form is achieved by combining the features of arch and beam to produce the stiffened tied arch. The Satsop River Bridges pictured in Figure 7.17 are of this type. Such structures have a fairly flexible parabolic arch rib and a stiff horizontal girder, which also acts as a tie for the arch thrust. A uniform load such as dead load is carried almost entirely by the arch, while a concentrated load induces bending in the girder in addition to arch compression and girder tie tension.

The arch of Figure 7.18 does not have the same dimensions as the Satsop River Bridges, but serves as a convenient example. An approximate

**Fig. 7.16** Gable Frame

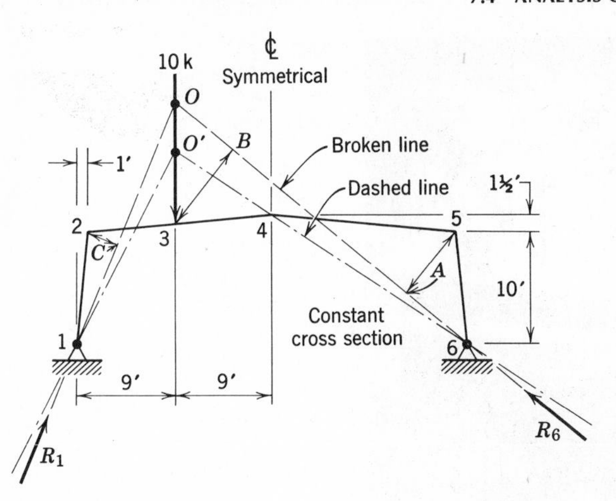


(*a*) Gable frame

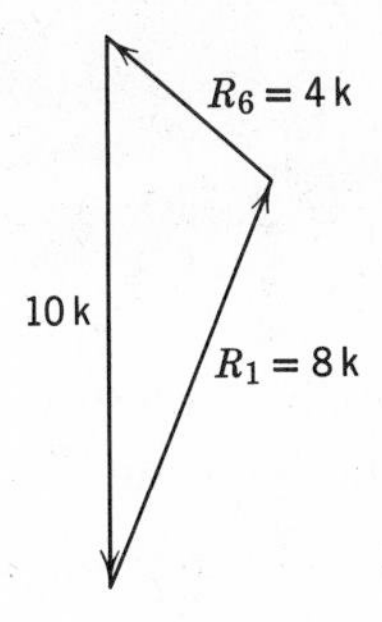

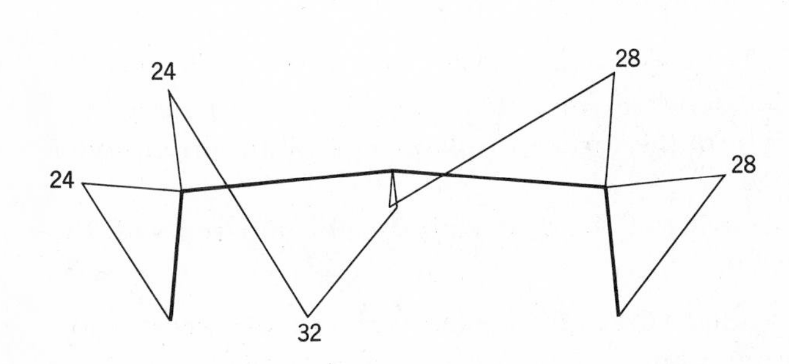


(*c*) Moment diagram, approximate analysis

*Garceau Steel Structures*

**Fig. 7.17**
Satstop River Bridges: 1968 AISC Prize Bridges Competition (short span), state of Washington.

*AISC*

analysis of this structure is to be performed. The analysis is based on the following behavior assumptions:

1. The arch rib will carry no bending moment since it is very flexible.
2. The hangers are inextensible since the elongation of the hangers is very small compared with the vertical displacements of the girder under the concentrated load.
3. The vertical displacements of the girder are small compared with the geometry of the bridge.

With these points in mind, the approximate analysis proceeds easily. Assumptions 2 and 3 imply that the arch rib remains parabolic in shape. The hanger loads are therefore equal, since a parabolic arch rib with no bending must be loaded uniformly. The hanger loads are assumed to carry the total applied load, and this gives $100/9 = 11.1$ k per hanger tension. The load on the arch is 11.1 kips at each panel or an equivalent uniform load of $11.1/10 = 1.1$ k/ft.

An arch in pure compression is readily analyzed. Equation 6.2 showed that the horizontal thrust is

$$H = qL^2/8h$$

where $q$ is the uniform load, $L$ the span, and $h$ the rise. In this example, $H = 1.1(100)^2/8(20) = 70$ k. The approximate girder tension and the horizontal component of arch compression is therefore 70 k.

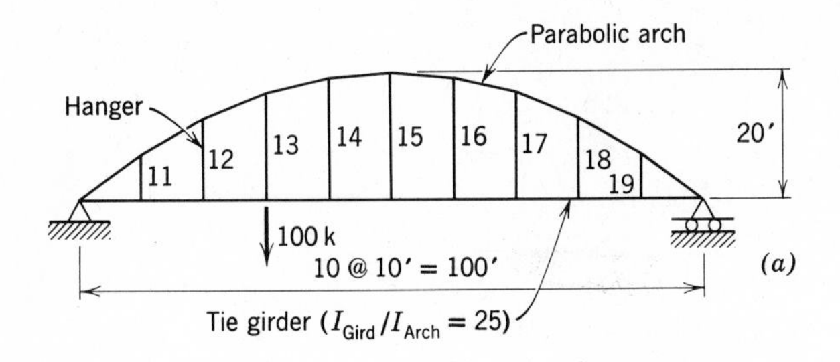


**Fig. 7.18**
Stiffened tied arch.

(*a*)

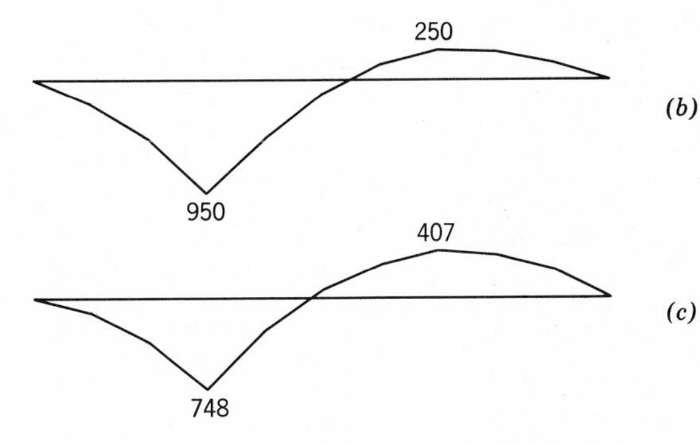


(*b*) Approximate girder moment (ft-k).

(*c*) Girder moment by conventional analysis (ft-k).

The bending moment diagram for the girder is found by considering the girder loaded with the applied load and the hanger loads. The resulting bending moment diagram is plotted in Figure 7.18*b*, with the moment diagram from an exact analysis plotted in Figure 7.18*c* for comparison. Table 7.3 indicates the hanger loads found by the exact analysis. The horizontal thrust by the exact analysis was 78.8 k.

**Table 7.3 Hanger Tensions from Exact Analysis of the Stiffened Tied Arch**

| Hanger | Tension (kips) |
|---|---|
| 11 | 11.8 |
| 12 | 12.2 |
| 13 | 15.7 |
| 14 | 12.3 |
| 15 | 12.1 |
| 16 | 12.1 |
| 17 | 12.1 |
| 18 | 12.1 |
| 19 | 12.0 |

## 7.5 SUMMARY

The study of approximate methods of analysis is difficult because of the variety of methods used, and the judgment needed to assess accuracy. Experience in structural behavior is the best basis upon which to develop this topic, but the use of approximate analysis aids in the generation of this experience. The beginning designer is urged to try these and other methods of approximate analysis on as many structures as possible and to examine results of exact analyses. You are also urged to try to generate your own methods on special structures that you may encounter. A gradual increase in accuracy of results and in understanding of structural behavior will evolve.

### *Suggested Reading*

Benjamin, J. R. [1959]: *Statically Indeterminate Structures,* McGraw-Hill, New York.
Norris, C. H., and Wilbur, J. B. [1960]: *Elementary Structural Analysis,* 2nd ed., McGraw-Hill, New York.

## PROBLEMS

7.1 Draw a bending moment diagram for the beams shown in Figure P7.1.

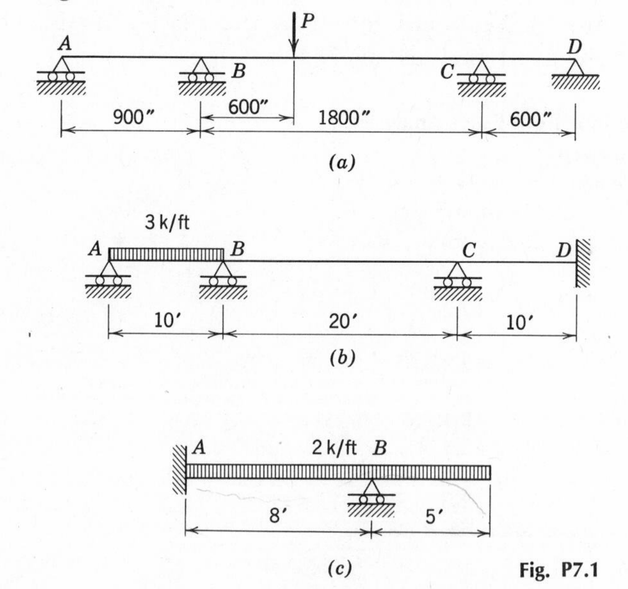


Fig. P7.1

7.2 A 20-k wind force is applied laterally to the mill building of Figure P7.2. Find the maximum column moment.

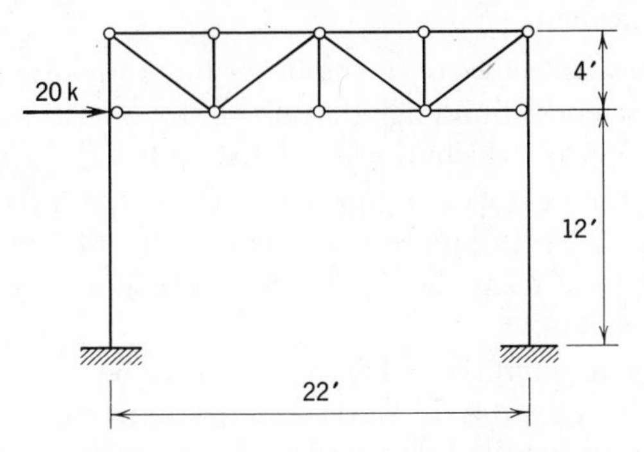


Fig. P7.2

7.3 A building contractor wants to build a temporary water supply tank from waterproof plywood and 2 by 6 beams. The vertical beams are in pairs and are spaced 2 ft apart along the entire length of the tank, as shown in Figure P7.3.

You are concerned with the bending moment in the vertical beams at some location near the center of the tank, where any end effects may be neglected. With two steel tie rods located as shown, determine the maximum bending stress in the vertical beams.

Is there a better location for the intermediate tie rod? For the upper tie rod?

Your solution should include a careful idealization of the

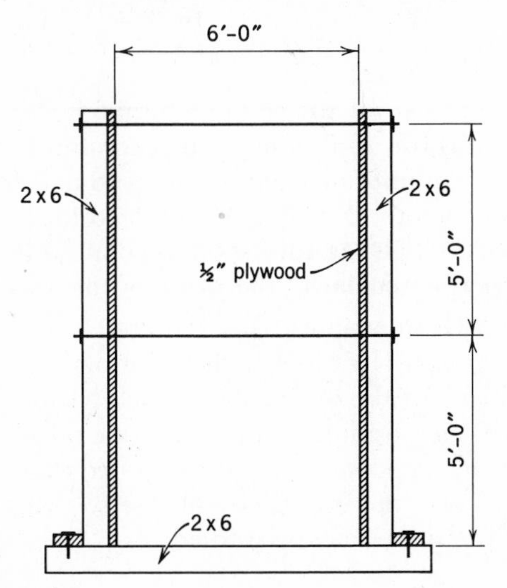


Fig. P7.3

structure, as well as justification for any assumptions you may wish to make in order to simplify the problem. An accuracy of ±10% is considered excellent.

7.4 Find the maximum column moment in the outer line of columns in the Cornell Social Sciences building. If you use the portal method, use the same shear distribution as in Example 7.7.

7.5 Use the portal method to perform an approximate analysis of the structure of Figure 7.13. Compare your results with the results of the cantilever analysis of Example 7.6 and with the exact results given with the example.

7.6 The concrete rigid frame building of Figure P7.6 is to be analyzed for its dead load condition. Devise an approximate analysis scheme to determine the important design quantities (axial forces and moments in beams and columns).
*Hint.* Note that symmetry prevents rotation of joints at the center column. Examine Figure 7.3 and try to find bounds on the location of beam inflection points.

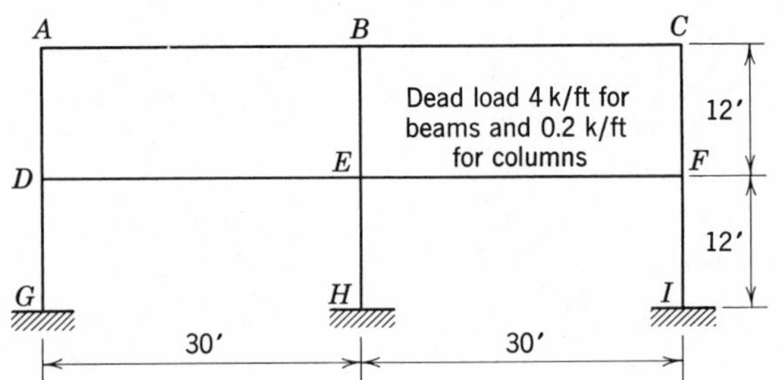


**Fig. P7.6**

7.7 The tall building frame proposed for the city of Chicago is shown in Figure P7.7. What is the approximate total column axial load expected at the base of an interior frame? Use wind and floor live load information from Chapter 3 and assume a dead load of about 100 psf of floor space. The frames are spaced at 30 ft centers in the dimension perpendicular to the plane of the figure.

7.8 Analyze the stiffened tied arch shown in Figure P7.8: (*a*) for a dead load of 4 k/ft and (*b*) for an HS 20-44 highway bridge loading. Assume that one lane of the bridge loads one arch. Your instructor will specify the position of the truck axle or of the concentrated load for the lane loading. It is suggested that each member of the class use a different value of these so that the effect of load type and position may be studied.

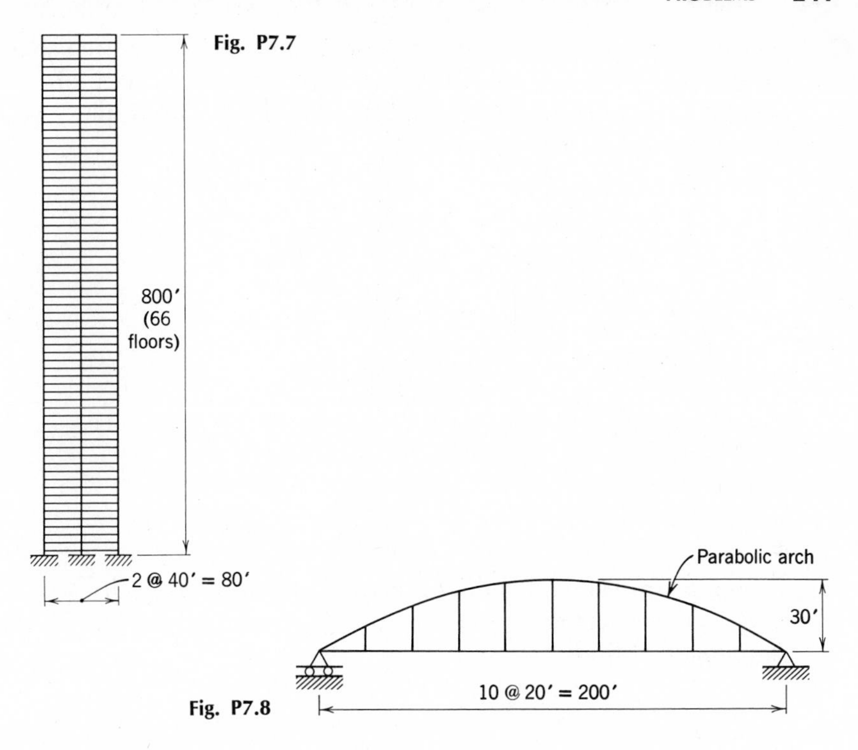


Fig. P7.7

Fig. P7.8

7.9 Determine the diagonal bar forces in the truss tower of Figure P7.9.
*Hint.* If you assume the diagonal bars are very slender so that they buckle elastically at low loads, you might neglect the force developed in the compression diagonals. Alternatively, if they are stocky, they may share the panel shear equally.

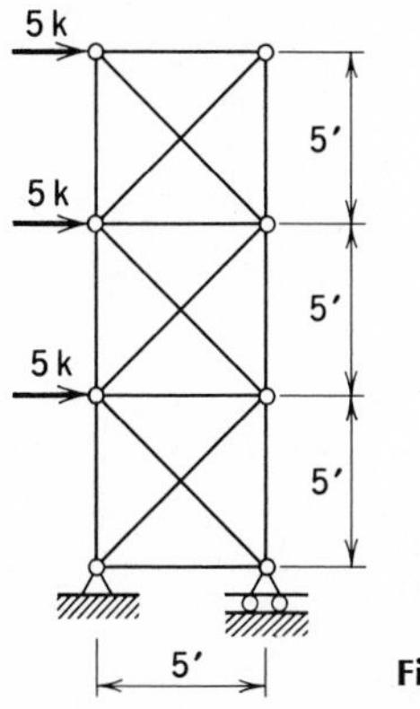


Fig. P7.9

7.10 Perform an approximate analysis, by sketching the deflected shape, of the frame shown in Figure P7.10. Also analyze for each of the loads separately and use superposition to get the result. Comment on the two approaches.

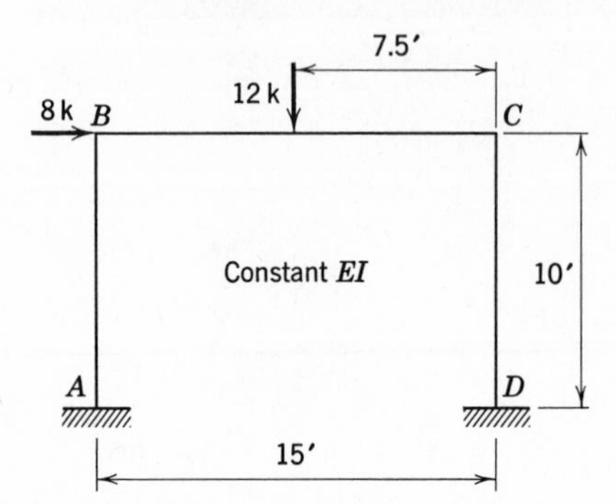


Fig. P7.10

7.11 Volume 2 of this text contains many examples of statically indeterminate structures. Try an approximate analysis for a number of these.

# References

# References


AASHO [1969]: *Standard Specifications for Highway Bridges,* American Association of State Highway Officials, 10th ed., Washington, D.C.

AISC [1970]: *Manual of Steel Construction,* American Institute of Steel Construction, 7th ed., New York.

Algermissen, S. T. [1969]: "Seismic Risk Studies in the United States," *Proc. Fourth World Conference on Earthquake Engineering,* Santiago, Chile, Vol. I.

ANSI [1955]: *Requirements for Minimum Design Loads in Buildings and Other Structures, A 58.1,* American National Standards Institute Code, National Bureau of Standards, Washington, D.C.

ASCE [1961]: "Wind Forces on Structures," Task Committee on Wind Forces, Committee on Loads and Stresses, Final Report, *Trans. ASCE,* Vol. 126, Part II, p. 1124.

Benjamin, J. R. [1959]: *Statically Indeterminate Structures,* McGraw-Hill, New York.

Bill, Max [1969]: *Robert Maillart—Bridges and Construction,* Praeger, New York.

Biswas, A. K., and Chatterjee, S., [1971]: "Dam Disasters: An Assessment," *Journal of the Engineering Institute of Canada,* March, Vol. 54/3.

BOCA [1970]: *The BOCA Basic Building Code,* Building Officials and Code Administrators International, 5th ed., Chicago.

Chinn, J., Mansouri, A. H., and Adams, S. F. [1969]: "Ponding of Liquids on Flat Roofs," *Journal of the Structural Division,* ASCE, Vol. 95, ST5, Proc. Paper 6539, May.

Cross, Hardy [1936]: "The Relation of Analysis to Structural Design", *Trans. ASCE,* Vol. 101, p. 1361.

Driver, G. P., and Miles, J. C. [1955]: *The Babylonian Laws,* Oxford Press, London.

Engel, Heinrich [1967]: *Structure Systems,* Deutsche Verlags-Anstalt GmbH, Stuttgart.

*Engineering News Record* [1940]: November 14, p. 649.

*Engineering News Record* [1941]: April 24, p. 589.

Galileo [1638]: *Dialogues Concerning Two New Sciences,* English translation by Henry Crew and Alfonso de Salvio, Dover Publications, New York.

Gies, Joseph [1963]: *Bridges and Men,* Doubleday, New York.

Golzé, A. R. [1971], "Model Law to Improve Dam Safety", *Civil Engineering,* Vol. 41, No. 3, March.

Holgate, H. J., Kerry, J. A., and Colbraith, J. [1908]: *Quebec Bridge Inquiry, Report of the Royal Commission,* printed by S. E. Dawson, Printer to the Kings Most Excellent Majesty (Queen's Printer), Ottawa.

Newmark, N. M. [1956]: "An Engineering Approach to Blast-Resistant Design," *Trans. ASCE,* Vol. 121, p. 45.

Norris, C. H. and Wilbur, J. B. [1960]: *Elementary Structural Analysis,* 2nd ed., McGraw-Hill, New York.

Quinn, A. D. [1961]: *Design and Construction of Ports and Marine Structures,* McGraw-Hill, New York.

SEAOC [1968]: *Recommended Lateral Force Requirements and Commentary,* Structural Engineers Association of California, San Francisco.

Sharpe, R. L., and Kost, G. [1971]: "Structural Response to Sonic Booms," *Journal of the Structural Division,* ASCE, Vol. 97, ST4, Proc. Paper 8063, April.

Steinman, David B., and Watson, Sara R. [1957]: *Bridges and Their Builders,* Dover Publications, New York.

Thom, H. C. S. [1968]: "New Distributions of Extreme Winds in the United States," *Journal of the Structural Division,* ASCE, Vol. 94, ST7, Proc. Paper 6038, July.

Timoshenko, Stephen P., and Young, Donovan H. [1965]: *Theory of Structures,* 2nd ed., McGraw-Hill, New York.

Todhunter, I., and Pearson, K. [1886]: *A History of the Theory of Elasticity and of the Strength of Materials,* Dover Publications, New York, 1960.

Waddell, J. A. L. [1916]: *Bridge Engineering,* Vol. I, Wiley, New York.

# Important Professional Organizations and Institutions

| | |
|---|---|
| AASHO | American Association of State Highway Officials |
| ACI | American Concrete Institute |
| AISC | American Institute of Steel Construction |
| AISI | American Iron and Steel Institute |
| AITC | American Institute of Timber Construction |
| ANSI | American National Standards Institute |
| ASCE | American Society of Civil Engineers |
| ASTM | American Society for Testing and Materials |
| BOCA | Building Officials and Code Administrators International |
| CRC | Column Research Council |
| PCA | Portland Cement Association |
| IABSE | International Association for Bridge and Structural Engineering |
| NBS | National Bureau of Standards |
| CRSI | Concrete Reinforcing Steel Institute |
| PCI | Prestressed Concrete Institute |
| SEAOC | Structural Engineers Association of California |
| WRC | Welding Research Council |

## ANSWERS TO SELECTED PROBLEMS

1.1 Note that roller supports are not always constructed from rollers; they also can be made with sliding plates. A hinged support may take on the shape of a rocker. Actual supports differ greatly in appearance from the idealized representations we use in structural line drawings.

2.1 Similarities include allowable stresses and elastic analysis. Differences occur in precision of load definition, possibility of testing, and permitting allowable stresses to be close to yield for spacecraft. Spacecraft are always checked for factored loads at ultimate stress conditions.

2.9 Note the large strain that occurs before rupture. Ductility is a vital characteristic of most structural steel.

2.10 Many building codes specify limits on drift. Values are in the range of $h/300$ to $h/500$.

2.13 Degree of uncertainty of the load and strength, accuracy of analysis, and consequences of failure all affect selection of factor of safety.

3.2 (a) In the Civil Engineering Building at Cornell University it ranges from 10 psf in a room seating one person per desk, to 16 psf in a fully occupied classroom with closely spaced fixed seats. (b) About 70 psf with dense crowding.

3.3 $d_0 = 6.08$ ft for a wind velocity of 100 mph.

3.6 The stress is 2580 psi compression in the aluminum and 1715 psi tension in the steel.

3.11 It is not safe to empty the pool; the minimum water level is 2.09 ft if we neglect friction.

4.1 There may be a measurable difference between calculated and experimental shapes if the weight of the chain or cord is significant in comparison to the applied loadings and if this weight is neglected in the calculations.

4.8 Possibilities for discussion include: (a) egg shell, (b) bone structure of a bird wing, (c) plant stem, (d) webbed foot of a duck, (e) tree structure, including trunk, limbs, and leaves.

4.9 Three greatly different animals are the dolphin, elephant, and antelope.

4.11 Possible structural solutions include: (a) shallow steel trusses (long span joists) spanning in one direction with rafters and roofing supported on the trusses; (b) several 60-ft steel or timber girders with shorter beams spanning between them; (c) reinforced concrete waffle slab spanning in both directions; (d) reinforced concrete or timber folded plate; (e) reinforced concrete or steel hyperbolic paraboloid.

4.14 *Hint.* The details of a highway bridge include the beam support mechanisms, beam connections, and expansion joints. These details represent a substantial part of the cost of a bridge, and your observations and comments should include discussion of these items and how they are influenced by the arrangement of spans in a particular bridge design.

5.1 (a) 0.96 in. and 1.65 in. Note lack of proportionality with load. (b) 0.023 in. and 0.046 in. Note proportionality with load. (c) Second line. Modulus $E$ used to compute $\epsilon_{Eq}$.

5.3 $H_A = 2.5$ k right, $V_A = 6.25$ k up, $H_G = 2.5$ k left, $V_G = 8.75$ k up

5.4 (a) stable, determinate, (b) stable, indeterminate to degree 1, (c) unstable, (d) unstable, (e) unstable, (f) unstable, (g) stable, determinate, compound truss, (h) stable, determinate, compound truss.

5.7 (a) stable, determinate, (b) stable, indeterminate to degree 2.

5.8 $n = 9$

6.2 (b) $H_A = 8$ k left
$V_A = 8$ k down
$V_B = 8$ k up
(c) $V_A = 3.73$ k up
$V_B = 10.67$ k up
(e) $H_A = 3.6$ k right
$V_A = 3.6$ k up
$R_B = 5.09$ k up and to left
(g) $V_A = 4.5$ k up
$V_B = 0.5$ k down
$H_C = 6.0$ k left
$V_C = 6.0$ k up
(i) $V_A = 1.07$ k up
$V_B = 0.53$ k up
$V_C = 4.27$ k up
$V_D = 2.13$ k up

6.3 $V_A = 1.30$ k down, $H_A = 3.12$ k left, $V_B = 1.30$ k up (per foot of structure)

6.4 (a) $\alpha = 14.5°$
(c) $\phi = 20.1°$

6.6 Wall is in equilibrium; horizontal friction force along $AB =$ 3.46 k vs 1.75 k soil pressure, and factor of safety against overturning = 13.20/5.84 = 2.26

6.8 $h = 1.75$ ft at center of right span; $T_{max} = 32.7$ k to left of middle support

6.10 Measuring sag values from the elevation of point $A$, $h_D = 17.44$ ft, $h_C = 23.54$ ft, $T_{max} = 22.3$ k at left end

6.12 (a) $F_1 = 28.3$ k tension (T)
$F_2 = 60.0$ k compression (C)
$F_3 = 8.0$ k (C)
(c) $F_1 = 0$
$F_2 = 7.5$ k (C)
(d) $F_2 = 11.9$ k (C)
(e) $F_1 = 14.4$ k (C)
$F_2 = 2.65$ k (C)
$F_3 = 0$

6.13 $F_{CF} = 4.38$ k (T)
$F_{AB} = 10.1$ k (C)

6.16 $M_D = -156$ ft-k (tension outside), $V = -2.2$ k, $F = 2.2$ k (tension).

6.18 5.25 k in the tension member and 2.11 k in each of the compression members.

6.19 $F_{57} = 63.0$ k (T), $F_{79} = 55.1$ k (T), $F_{47} = -10.2$ k (C), $F_{67} = 15.9$ k (T).

6.21 (c) $M_B = -865$ ft-k, $M_C = -257$ ft-k, $M_D = 780$ ft-k.

6.22 (c) $M_B = 12$ ft-k, $M_E = -55$ ft-k.

6.25 $M_B = M_D = 6.85$ ft-k, $M_C = 7.4$ ft-k.

6.28 $M = -0.436PR$

6.30 $M_{12} = 67.5$ ft-k, $M_{23} = 22.5$ ft-k, $M_{25} = 9.0$ ft-k, $M_{36} = 22.5$ ft-k.

6.32 (a) $D = 5pL^4/384EI$, (b) $D = PL^3/3EI$

7.1 (a) elastic analysis: $M_B = -12.8P$ ft-k
$M_C = -4.8P$ ft-k
(b) elastic analysis: $M_B = -14.2$ ft-k, $M_C = 5.2$ ft-k, $M_D = -2.6$ ft-k.
(c) elastic analysis: $M_A = -3.53$ ft-k, $M_B = -25.0$ ft-k.

7.2 Portal method: 7.2 ft-k.

7.4 Cantilever method: 46.2 ft-k middle column,
Portal method: 34.2 ft-k middle column, Elastic analysis: 22.1 ft-k lower column.

7.5 Compare with Table 7.2. Portal method gives:

| Member | Axial | Shear | Moment |
|---|---|---|---|
| 1 | −9.0 | −1.5 | 15 |
| 2 | −3.0 | −2.5 | 15 |
| 3 | −7.5 | −4.2 | 42 |
| 4 | −2.5 | −7.1 | 42 |
| 5 | −7.5 | −6.7 | 67 |
| 6 | −2.5 | −11.2 | 67 |
| 7 | −7.5 | −9.2 | 92 |
| 8 | −2.5 | −15.4 | 92 |
| 9 | −3.8 | −17.0 | 170 |
| 10 | −1.3 | −28.3 | 170 |
| 11 | 38.8 | 11.8 | −118 |
| 12 | 25.8 | 23.5 | −235 |
| 13 | 64.6 | 11.8 | −118 |
| 14 | 21.8 | 10.5 | −52 |
| 15 | 14.5 | 21.0 | −105 |
| 16 | −36.2 | 10.5 | −52 |
| 17 | 12.5 | 8.0 | −40 |
| 18 | 8.3 | 16.0 | −80 |
| 19 | −20.8 | 8.0 | −40 |
| 20 | 5.8 | 5.5 | −27 |
| 21 | 3.8 | 11.0 | −55 |
| 22 | −9.6 | 5.5 | −27 |
| 23 | 1.5 | 3.0 | −15 |
| 24 | 1.0 | 6.0 | −30 |
| 25 | −2.5 | 3.0 | −15 |

7.6 Authors' approximate analysis:
Top beam: $M^-_{max} = 392$ ft-k, $M^+_{max} = 207$ ft-k, $F = 16.3$ k (C).
Lower beam: $M^-_{max} = 340$ ft-k, $M^+_{max} = 169$ ft-k, $F = 2.3$ k (C).
Column *AD:* $M_{max} = 115$ ft-k at top, $F = 53.2$ k (C).
Column *DG:* $M_{max} = 150$ ft-k at top, $F = 112$ k (C).
Column *BE*: $F = 141$ k (C), Column *EH*: $F = 270$ k (C).
Columns *BE* and *EH* have no moment because of symmetry.

7.9 Assuming that compression members buckle, the lower bar sloping up to the right carries 21.2 k (T). Assuming that all diagonals are stocky, the lower panel bar sloping up to the right carries 10.6 k (T) and the other diagonal has 10.6 k (C). Either assumption can be satisfied by proper selection of the cross section of the diagonals. Note that diagonal rods are often pretensioned to avoid buckling when they are loaded in compression; thus both diagonals in a panel participate in carrying load.

## INDEX

The first number for each entry is the primary reference.